Electronic Circuit Cards and Surface Mount Technology

A Guide to their Design, Assembly, and Application

To my son, Christopher George
and daughter, Kathleen Gena

Electronic Circuit Cards and Surface Mount Technology

A Guide to their Design, Assembly, and Application

Malcolm R. Haskard

Director and Associate Professor
Microelectronics Centre
University of South Australia

PRENTICE HALL

New York London Toronto Sydney Tokyo Singapore

Acquisitions Editor: Andrew Binnie.
Production Editor: Fiona Marcar & Fiona Henderson.
Cover design: The Modern Art Production Group, Melbourne, Victoria.
Typeset by Midnight Express, Cromer, NSW.

Printed in Australia by Impact Printing, Brunswick, Victoria.

1 2 3 4 5 96 95 94 93 92

ISBN 0 13 249988 6.

National Library of Australia
Cataloguing-in-Publication Data

Haskard, M. R. (Malcolm Rosswyn).
Electronic circuit cards and surface mount technology.

Includes bibliographies and index.
ISBN 0 13 249988 6.
1. Printed circuits - Design and construction.
2. Electronic circuit design. 3. Electronic circuits.
4. Surface mount technology. I. Title.

621.381531

Library of Congress
Cataloging-in-Publication Data

Haskard, M. R. (Malcolm R.), 1936-
Electronic circuit cards and surface mount technology: a guide to their design, assembly, and application / Malcolm R. Haskard.

p. cm.
Includes bibliographical references and index.
ISBN 0-13-249988-6
1. Printed circuits--Design and construction. I. Title.

TK7868.P7H38 92-26810
621.3815'31--dc20 CIP

Prentice Hall, Inc., *Englewood Cliffs, New Jersey*
Prentice Hall Canada, Inc., *Toronto*
Prentice Hall Hispanoamericana, SA, *Mexico*
Prentice Hall of India Private Ltd, *New Delhi*
Prentice Hall International, Inc., *London*
Prentice Hall of Japan, Inc., *Tokyo*
Prentice Hall of Southeast Asia Pty Ltd, *Singapore*
Editora Prentice Hall do Brasil Ltda, *Rio de Janeiro*

PRENTICE HALL

A division of Simon & Schuster

Foreword

A book on Electronic Circuit Cards is timely as the technologies for realizing essential elements of electronics equipment using such assemblies have come of age.

While there were many doubting Thomases about Surface Mounting Technology (SMT) because of the many problems caused by the integrity of both the electrical and mechanical connections being dependent on the solder joint, as the dominant component assembly technology of the 1990s SMT is here to stay. I recollect attending a prominent European conference during the early 1980s where a speaker from Hamburg forecast that SMT was doomed and that everyone should stick to the old, well-tried, through-hole mounting technologies. This gloomy outlook ran counter to the more informed and well-researched technical and market predictions, that new technologies were being created for both technical and commercial reasons. The dangers of the speaker's lack of vision are now evident. Fortunately, he was not in a position to do significant harm and intelligent members of the industry proceeded to adopt the technologies necessary to meet the challenges of the future.

The point of the anecdote is that it is necessary to anticipate and be prepared to take on new technologies, and that the effective foil to the pessimists is to be constructive and to build for the future. One essential route is to develop effective training programs and that is exactly what Malcolm Haskard and his colleagues are doing at the Microelectronics Centre, which is now part of the University of South Australia. I have great respect for the Centre's work in information gathering from around the world and intelligent application of that knowledge to establish practical facilities and a sound base of information.

Malcolm Haskard has now converted that sound base of information into a coherent text to provide a solid grounding for students and professionals to learn about techniques of designing and realizing today's electronics assemblies. He deals with the range of subjects in 11 well-constructed chapters, ranging from the basic assembly technologies through to PCB technology, soldering, alternative joining techniques, added components, use of screen printing for paste or adhesive application, and the actual assembly inspection and testing. He goes on to provide an insight into production control and approaches to design, and closes with a chapter on related techniques and components, including a discussion of a field in which he has made significant contributions—sensors.

I am pleased to have had the opportunity to write this Foreword for a valuable education text published by an able teacher who has become a friend, and also to recommend the text to those wishing to bring themselves up to speed about the range of technologies that create Electronic Circuit Cards.

Nihal Sinnadurai
Ipswich
England
June 1992

Contents

Preface

The printed circuit card has made such an impact on the electronics industry that it could be argued it is one of the most important inventions this century. The idea was patented in the United Kingdom in February 1943, but it was not until the late 1950s that industry started to appreciate and use this 'new' technology. Today it is difficult to find an electronic product that does not use a card in some shape or form. Simplistic single-sided cards are used in toys; miniature cards in watches and hearing aids; racks of double-sided plated through-hole cards in the telecommunications industry; compact multilayer cards in computers and aircraft navigational equipment; and flexible cards under dashboards of cars are a few examples.

Another revolution is taking place as a result of surface mount technology. Smaller, densely packed cards that can be produced in large volumes are appearing, and technicians, service personnel and engineers alike must prepare for this new generation of cards. The technology is already developing new ideas like molded cards.

For 12 years the Microelectronics Centre at the University of South Australia has been providing training in this area. This experience has revealed the need for an appropriate textbook. Often at the trade and certificate levels soldering skills are taught in isolation, and the total picture of the industry is not presented. At the other extreme, professional engineers are frequently not provided with practical information, such as wire types and coil-winding methods.

Electronic Circuit Cards brings all this material together, providing a valuable source for a range of courses at various levels, be it trade, certificate, paraprofessional or professional engineering. At the elementary level the book can be used for teaching the basic skills of hand soldering (the appendices include a number of practice board designs and projects). At the more advanced level it provides an introduction to essential information on the total electronic manufacturing process, including electronic components, card design, card manufacture and assembly (both manual and automatic), fundamentals for establishing production lines and testing.

An explanation is needed for the use of the term 'circuit card' rather than the more common 'circuit board'. The latter tends to be restricted to laminate materials and because this book endeavours to cover a range of materials, including ceramic, silicon and plastics, the more general term of 'card' is more accurate. The term 'circuit board' is used when appropriate, however.

I would like to thank those who have given valuable assistance, particularly H. T. (Bill) Bilske and John Duval from the School of Electronic Engineering, Barry Playford from Tekpro (Teknis), Dr Nihal Sinnadurai of British Telecom, and Grant Pearce, who introduced me to the finer points of high reliability hand soldering.

Also my thanks to Alan Marriage. Together we established the Surface Mount Unit at the Centre. My special thanks to John Banbury for taking many of the photographs; to Isobel Keegan for typing the draft of this text; and last, but not least, John Crawford of Philips Components, whose support, comments and advice on the draft were much appreciated.

Finally, to both teacher and student, I trust that you will find *Electronic Circuit Cards* both informative and easy to read. Above all, I hope it will arm you with essential knowledge to face the electronic card revolution.

Malcolm R. Haskard

1 Introduction to circuit cards

1.1 An overview

The physical assembly of electronic components into a system to achieve reliable and repeatable electrical characteristics, good mechanical strength and a pleasant appearance has always been an essential objective in the electronics industry. Up until the 1950s, systems were assembled using discrete wiring with components mounted on tag strips and sockets. In low-cost consumer products the tag strips and sockets were made from synthetic resin bonded paper (SRBP), bakelite, and plastics, while for professional high-quality products molded ceramic was preferred (see Figure 1.1).

In response to the growing pressures of repeatable performance, miniaturization and lower costs, many alternative assembly schemes were developed and tried. One that proved very successful was the printed circuit board/card (or printed wiring board). It consisted of a laminate made from an insulating material like SRBP and was typically 1⁄16 inch (1.6 mm) thick. Onto one side was fixed a layer of copper foil. This foil was

Figure 1.1 Example of a high-quality ceramic tag strip

selectively etched to form the desired wiring pattern, holes were drilled through the laminate so that standard leaded components could be held to the non-copper side of the board, and leads were passed through the laminate and soldered to the copper (see Figure 1.2). The process had the potential to be automated and created a product that was both robust and gave repeatable electrical performance.

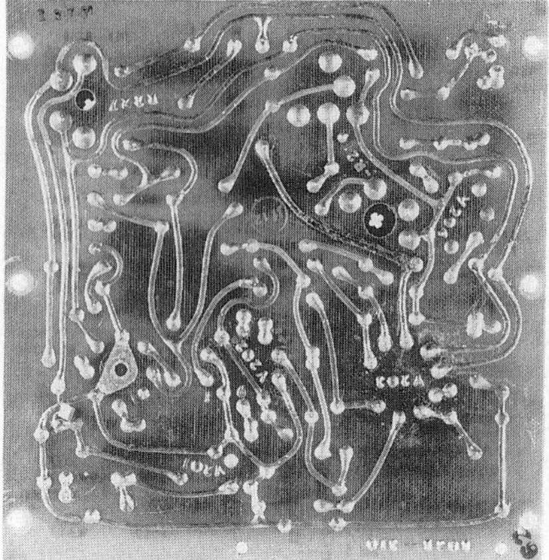

Figure 1.2 Example of an early printed circuit board

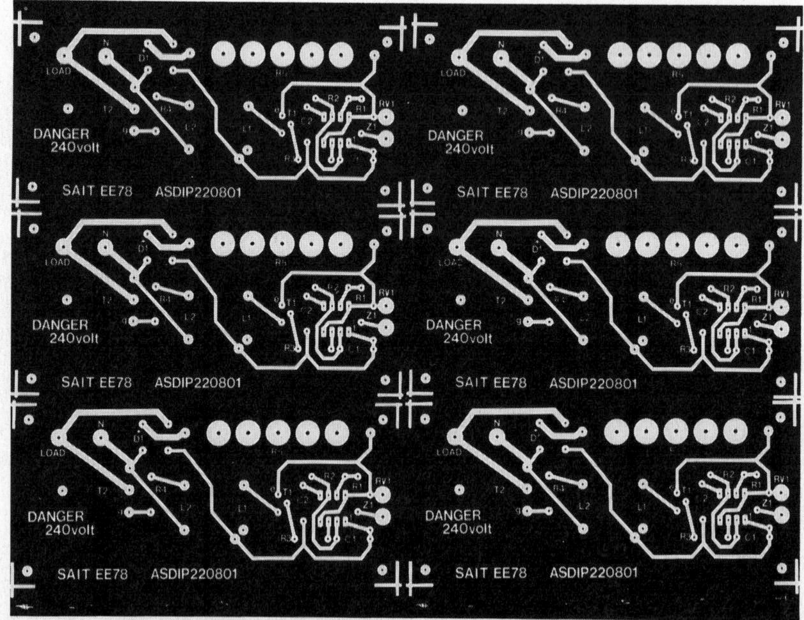

Figure 1.3 Example of a negative containing a matrix of boards

An alternative manufacturing method to 'glueing and etching' the copper to the insulating board was to plate the copper onto the board where a track was needed. It was a more expensive process, however, and while it is still occasionally used for fine line circuits, it did not 'catch on'. Research continued to improve the quality of printed circuit board and make it more suitable for mass production. Board materials that could be stamped or pressed, or provide greater strength were used. Flow soldering methods were introduced that allowed all components to be soldered in a single operation. This meant that boards with superior high-temperature properties were needed so that during soldering they did not bow or fracture. The boards must neither absorb the chemicals used for the etching of the copper nor react with them. In the early years, the etchant was usually ferric chloride, a corrosive material that could also leave a rusty orange stain.

To define the areas of the copper to be selectively etched, two methods were commonly used. The first employed a screen printing process, with the lacquer being printed onto the unetched copper side of the board, protecting the copper by covering it and being resistant to the etchant. After etching, the lacquer was stripped off. The second method employed photolithography, where the copper was covered with a photoresist material and, once dried, exposed to ultraviolet light through a photographic negative containing the wiring track pattern. The exposed photoresist polymerises and, when developed, does not wash away, but remains covering and protecting the copper underneath like the lacquer in the screening process. To speed up the production of cards, step and repeat methods were employed so that both the screen used for printing and the negative employed in the photolithographic process consisted of a matrix of board patterns. The process produced a matrix of boards which was later either sawn or punched into the individual elements (see Figure 1.3). These elements were sometimes called 'biscuits'.

The potential of this new electronic assembly process was exploited. Double-sided boards, that is, with copper on both sides and plated through holes to connect one side to the other, appeared. Flexible insulating material such as polyester with copper foil were used to construct wiring harnesses (see Figure 1.4a). Nickel plating the copper (the nickel could be used as the mask to selectively etch) allowed microwelding of components to the board for increased packing density. In more recent examples the boards or laminates consist of many alternate layers of insulator and etched copper conductors, called multilayer boards (see Figure 1.4b). Special copper layers can be used for screening and power supply rails while other materials may be added to help dissipate heat or adjust the temperature coefficient of expansion of the board to match that of the components mounted on it. The printed circuit board has proved to be a versatile, reliable method of assembling electronic systems.

Two methods which were developed to increase the packing density of components in a system were the 'Cordwood' and the 'mother-daughter' board systems. They both make use of planar printed circuit boards, configuring them into three-dimensional systems. Figure 1.5 illustrates the techniques.

In the case of the Cordwood construction, two cards are used with axial components between them. For connection between the two cards, a wire link is used, inserted as any other component. The difficulty with this system is that components in the center have poor ventilation and, if they fail, are difficult to replace.

The mother-daughter board system allows small cards, with connections along one edge (single in line) called 'daughter boards', to be mounted vertically on a second larger board called the 'mother board'. It is still a popular method. Daughter boards can be made using other techniques, such as thick and thin film. An extension of this concept is the back plane in a rack of equipment. Printed circuit cards in slides plug into a socket at the rear of the equipment, the sockets being mounted on another printed circuit card. This latter card has all the interconnection wiring between sockets and any additional components such as power supply bypass capacitors.

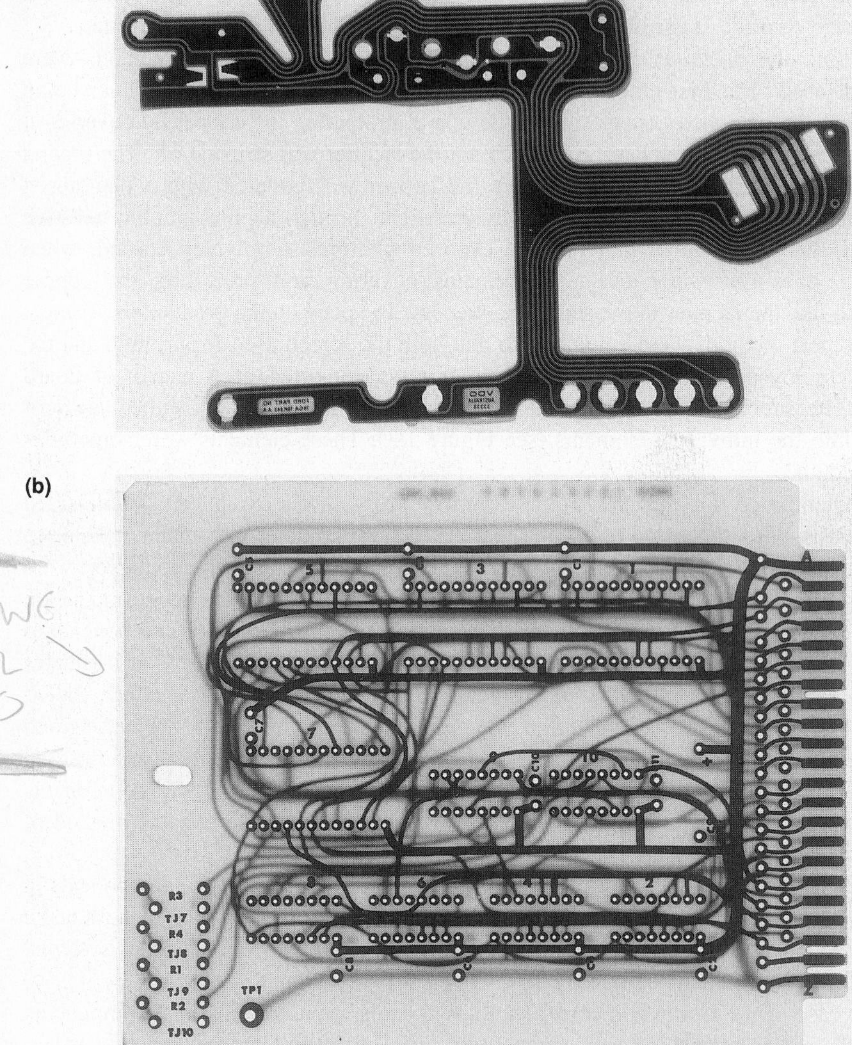

Figure 1.4 Examples of different board construction: (a) flexible, and (b) multilayer. (Note that the embedded tracks can be seen when the board is backlit)

(a)

(b)

Figure 1.5 Examples of: (a) Cordwood and (b) mother-daughter board construction
(Reproduced with permission of Grant Pearce and Philips Components)

1.2 Related technologies

Following the success of the printed circuit board, there were bound to be adaptations of it. Two are frequently used today. The first is the wire-wrap method devised by Bell

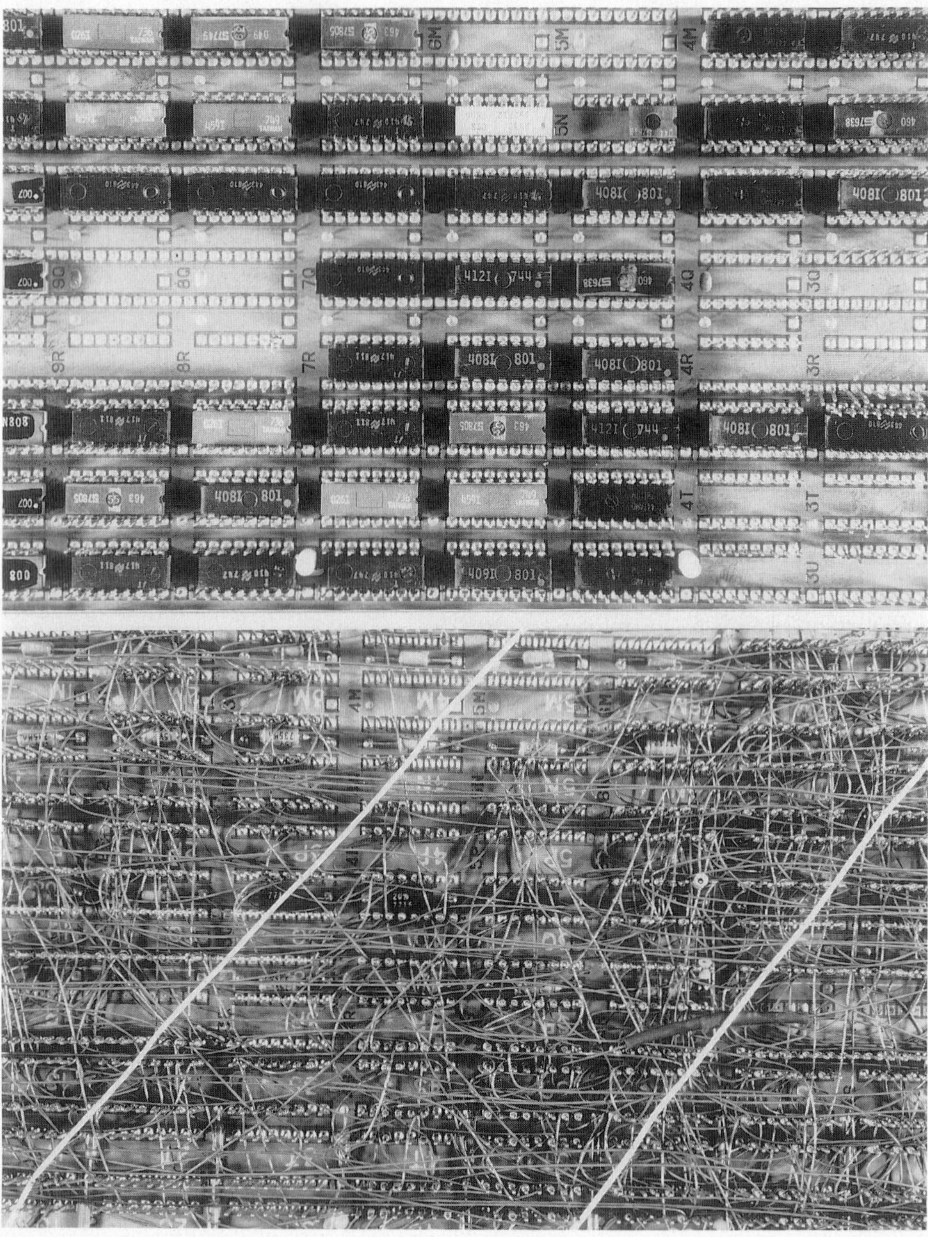

Figure 1.6 Example of a wire-wrap construction

Telephone Laboratories in the United States for use in telephone exchanges. It is illustrated in Figure 1.6 and discussed in detail in Section 4.4.

Pins are extended from the copper side of the card and the interconnecting wires are wrapped around the pins with a special tool. The same tool can unwrap and carry out repairs. No soldering is required. While the original system employed pins on an 0.2 inch (5.08 mm) grid, the present system uses the standard 0.1 inch (2.54 mm) grid and components normally plug into dual in line (DIL) sockets, which are attached to the pins. The function of the printed circuit card is to provide a mechanical base for the array of sockets, distribute power to the sockets and hold down any non-DIL components such as large electrolytic capacitors. Occasionally sockets are not used; the DIL packages are inverted and soldered to the wire wrap pins protruding through the component side of the card (see Figure 1.6).

Another common application of wire-wrap is in the area of prototyping and experimentation. For example, many microprocessor kits have the basic component of the kit assembled as a standard printed circuit board card which has a section pre-drilled and left for wire wrapping. This allows purchasers to add other components and adapt the kit to their particular needs.

The second adaptation is the multiwire, multilayer system. As the packing density on a printed circuit card increases, the number of conducting layers in the card must also increase. Seraphin (1978) examines the length of interconnecting signal wires between integrated circuits on a card and estimates that the length of wire per package site L is given by:

$$L = 1.125\, SP \qquad [1.1]$$

where S is the distance between adjacent packages
P is the number of pins per package

Thus, for example, if S is 0.5 inch and P 68, the length of wire per package site is over 38 inches.

Cards of ten layers are not uncommon and the cost of tooling to produce them in small quantities is prohibitive. The multiwire system allows the production of small quantities of boards, equivalent to multilayer boards, at reasonable costs.

A double-sided printed circuit card is produced and then fine insulated copper threads are welded to the board to build up the system. The wire is typically 160 micron in diameter. When the wiring is completed the threads are buried in a layer of epoxy and cured. The double-sided board, with the added wire, can be equivalent to 10 or more conventional multilayers. An improved process called 'microwire' is now available in which the insulated wire has been reduced to 100 micron in diameter and the production of up to 60 equivalent multilayers is possible.

1.3 Alternative technologies

Although the printed circuit card technology is probably the most common assembly method in the electronics industry, it is not the only one. The two most important alternative technologies are film and monolithic. There are two types of film processes. The first, thick film hybrid (Haskard 1988) is used extensively when high reliability, high temperature, robustness or compactness are required at relatively low cost. The process is based on sequential screening of pastes (which may be conductive, resistive or dielectric) onto an insulating substrate with firings in between each printing to cure the pastes (see also Chapter 6). In this way passive multilayer circuits can be built up (see Figure 1.7). Active components must be added later, either in packaged or unpackaged form. The substrate material is normally alumina, but families of low-temperature silver-based pastes allow screen printing on plastics, including printed circuit board material (see Section 6.3). The pastes employed are either based on expensive noble metals such as gold, silver, or platinum or base metals like copper, chromium, or nickel. The latter paste types have not proved popular and are still considered to be in the research phase. The printing tolerance for components is typically ± 20%, but both active and passive trimming allows component tolerances to be made 1% or better, and circuit parameters, like offset voltage of an amplifier, to be reduced to specific limits.

The second film process is thin film hybrid which uses vacuum techniques such as evaporation and sputtering to deposit conducting and insulating layers (Maissel & Glang 1970; Coutts 1978). Often processes such as electroplating and anodising supplement these basic vacuum steps. Two systems are commonly employed: the first and simplest uses gold for conductors, nichrome for resistors and silicon monoxide for a dielectric; the second employs tantalum for resistors and its oxide for dielectric. Normally, active components have to be added. The process is capable of manufacturing short-life, active devices such as cadmium sulphide field effect transistors (short life in that the characteristics degenerate quickly in a matter of a few days). While the cost of producing thin film circuits generally costs five times more than thick film, the quality of the finished product is often superior. Fineness of lines, resistance tolerance and stability are evidence of thin film's superiority to thick film.

The monolithic technologies, principally silicon and, to a lesser extent, gallium arsenide application specific integrated circuits (see Figure 1.8), are not directly

Figure 1.7 Example of a thick film circuit

competing with printed circuit board technology. Most manufacturers use printed circuit boards and film technologies to interconnect monolithic devices. This may not always be the case. Just as it took several years for manufacturers to realize that processing discrete transistors on a wafer, sawing them up and then reconnecting them to form a circuit was unnecessary, costly and reduced system reliability, the same is being discovered about packaging chips and mounting them on cards or substrates. Wafer-scale integration is one method which involves adding extra metalisation layers to interconnect good dies on a wafer together to form a system (Jesshope & Moore, 1986). The most common example is the large random access memory (RAM). Silicon wafers are also starting to be employed as substrates in a similar way to printed circuit boards. The same technology for producing the chip can be used to form the fine interconnection tracks between dies on a blank wafer. Dies may be inverted and mounted onto the wafer surface or, alternatively, mounted with the active surface upwards in etched grooves or holes in the silicon wafer. Since both dies and substrate are of the same material, the disadvantage of having materials of differing thermal coefficients of expansion is eliminated (Dettmer 1988).

Yet another way of assembling several dies to form a system is the multichip module, which uses multilayer thick or thin film substrates to achieve the interconnection (Geschwind & Clary 1991). Various forms exist and for some years now the concept has been used to produce memory modules. The rapid increase in clock frequencies, very large pin count silicon dies, coupled with an established, reliable, high packing density, low cost technology has created widespread interest in this method of assembly.

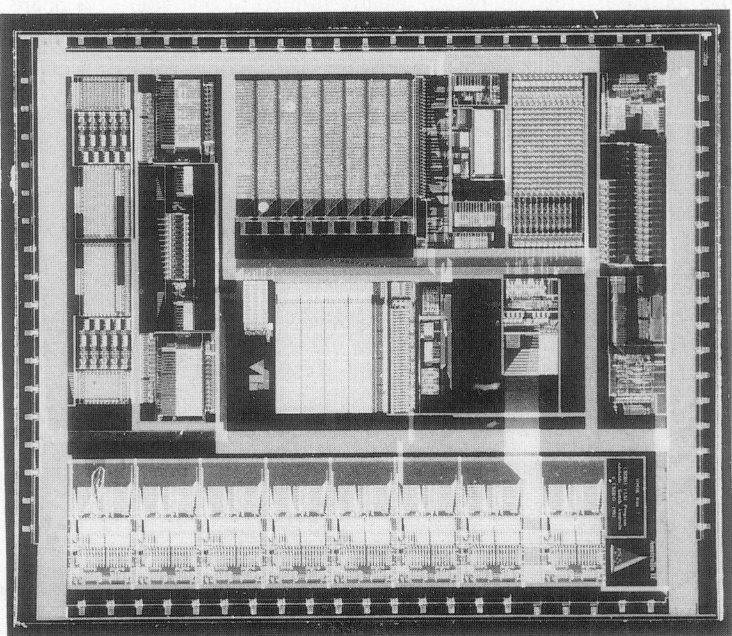

Figure 1.8 A VLSI chip of 100,000 transistors

With complete systems on a single chip there is no reason why the chip package cannot contain all external components and a plug for interconnecting to the system. Standard DIL packages with built in bypass capacitors have been available for some years so the single chip package system cannot be too far away. Figure 1.9 summarizes the competing technologies as a tree diagram.

1.4 Surface mount technology

The rapid growth in very large scale integration (VLSI) technology, where chips with 64 pins or more are becoming commonplace, coupled with the size advantage of film circuits, has forced a significant change in printed circuit board methods, namely the change to surface mount technology (SMT). A problem with conventional printed circuit board methods is that the plated-through hole is employed for mounting leaded components as well as making interconnections between layers. The result is a plated-through hole that is physically larger than one simply used as a via. Surface mounted or leadless components allow the components to be mounted on the surface, as is the case for film hybrids, and therefore a size reduction is possible. With the elimination of holes for mounting components, components can now be mounted on both sides of the boards and leads centered on pitches less than 100 mil (2.54 mm), typically 50, 40 and even 25 mil. Fine line techniques must now be used on printed circuit boards.

The elimination of holes to mount components might suggest that there will be less holes in a board, allowing a cost reduction. In practice this does not seem to happen. With the reduction in board area, multilayer boards become more frequent and the number of holes for vias increases. Test points are usually limited to one side of a board and this further increases the number of vias, as leads must be brought through to this side from the specified test points. A more recent innovation is to mount MELF (Metal ELectrode

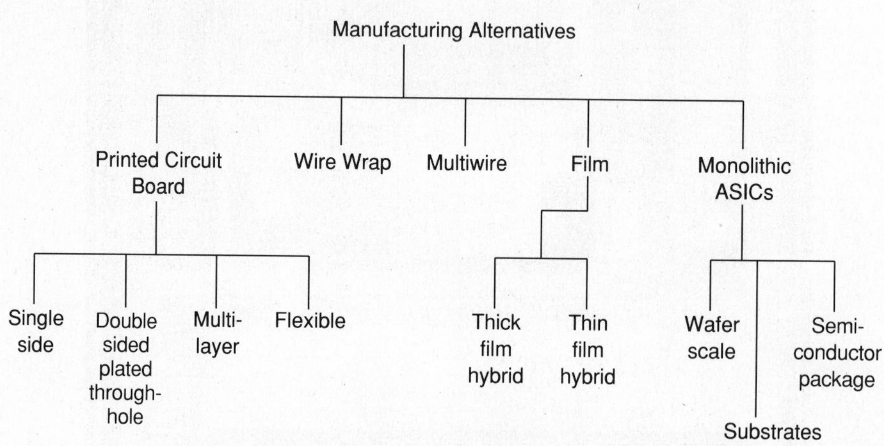

Figure 1.9 Various manufacturing alternatives

Face bonded) surface mount components vertically in holes in the board, thereby reducing the size of a board.

Because surface mount technology was the standard procedure for film hybrids, it was initially thought that its introduction to printed circuit board technology would be a simple matter. There were hidden problems, however. First, many of the larger VLSI chips used leadless ceramic packages. This did not matter in the case of the film technologies because the substrates were also ceramic and there was, therefore, a matching of the coefficient of thermal expansion. Such is not the case for printed circuit

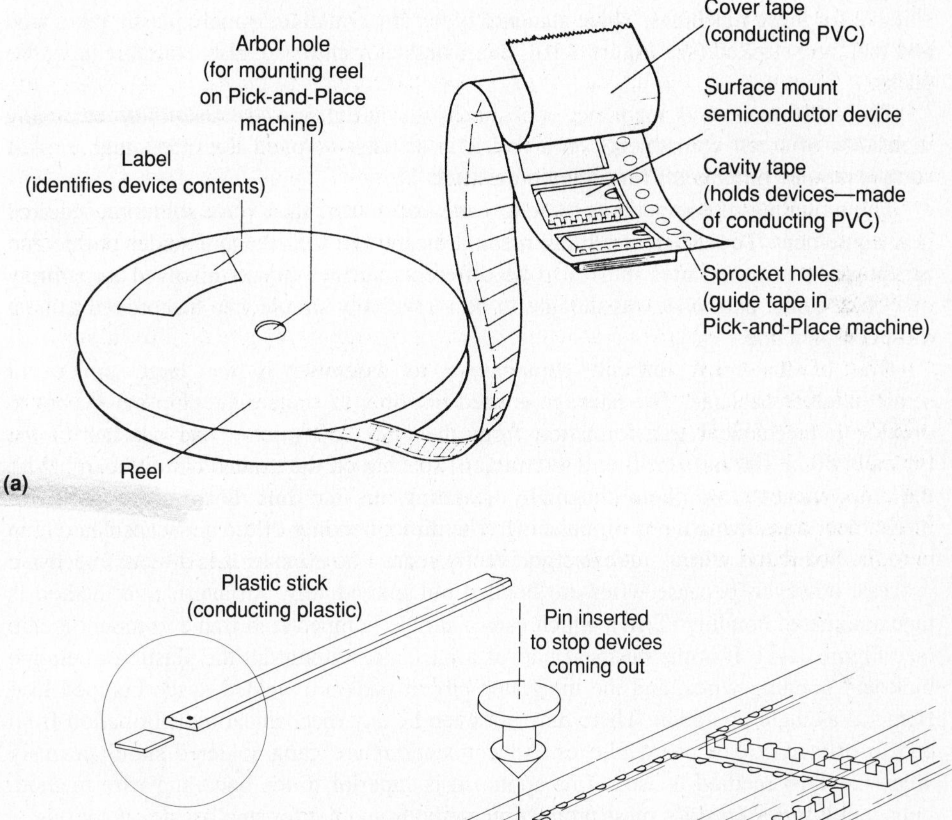

Figure 1.10 Packaging of surface mount components: (a) tape and reel, and (b) plastic tube

boards, which normally are made of an epoxy material. Matching of thermal coefficients by using correct fillers, flexible surfaces and including special metal layers in the laminates were possible solutions. Some, such as the inclusion of a layer of sheet invar are very expensive. A common solution now is to use leaded surface mount integrated circuit components, usually gull wings and J leads.

Even the supply of surface mount components was an unexpected problem. The change over from leaded to surface mount types has taken longer than anticipated. Further, because surface mounted technology is ideal for mass production using pick-and-place machines to load boards, the components are needed in carriers that are suitable for these machines. Three standard types are available, namely plastic tube, tape and reel, and stacked (see Figure 1.10). Semiconductor chips are also available in waffle packs.

The problem now is the exact opposite of the initial one. Because there are many thousands of small components on a reel it is difficult to purchase small quantities of surface mount components for prototyping work.

The mounting of components by reflow methods rather than wave soldering required new equipment. The different joining method meant that the resultant solder profile and acceptance test procedures must also be different. Surface mount involved an entirely new 'ball game' and what was thought to be a relatively simple step has proved to be a most complex one.

One of the most difficult components to assemble is the large pin count semiconductor package. The package is used not only to protect the chip inside, but to provide a mechanical transformation from the chip 250 micron pad spacing to the typically 50 or 100 mil (1250 or 2500 micron) spacing on the printed circuit board. With the improvements in photolithography resulting in fine line boards, the need for mechanical transformation is diminishing. The direct bonding of the unencapsulated chip onto the board and wiring out can significantly reduce board size. It is a more expensive process, however, because wires are bonded out individually. An alternative method is tape automated bonding (TAB), which uses a simple bumped lead frame to mount a chip (see Figure 1.11). It is the business part of a package, without all the plastic or ceramic bulk and bonding wires, and the integrated circuit pads are welded to the bumped lead frame in a single operation. There may not even be any mechanical transformation from pad to printed circuit board pitches. All connections are gang soldered simultaneously when the TAB method is used. This approach is superior to the bond and wire method, but again it is designed for mass production methods and not for small-scale operations.

Thus the merging of surface mount and printed circuit board technologies has much to commend it, providing improved automatic assembly and smaller, less expensive, products. There are, however, numerous problems which require further research and the development of new processing methods.

1.5 Comparison of the manufacturing technologies

The method employed to manufacture a system must be taken into consideration during the design phase. Important factors include cost, maturity of the technology, size, power dissipation and frequency of operation of the finished product, required reliability, maintenance and repair, security of the product, volume of the product, and time available to complete the job. Table 1.1 gives in summary form a comparison of the major manufacturing technologies. Wafer-scale integration and wafer substrates have not been included because they are still in the development stage.

Figure 1.12 shows that application specific integrated circuits have the greatest potential for producing the lowest cost product even in small volumes. But many systems either do not require such complexity or are restricted by needing a card of some form to interconnect several integrated circuits or other devices such as relays, opto isolators, transformers and connectors. While printed circuit boards will never compete with the complexity of application specific integrated circuits, the introduction of surface mount and other modern techniques is reducing the cost per connection and increasing the number of connections possible, so that the position of the multilayer printed circuit board/hybrid in Figure 1.12 is slowly moving in the direction of the CMOS application specific integrated circuit.

Figure 1.11 Example of a TAB chip

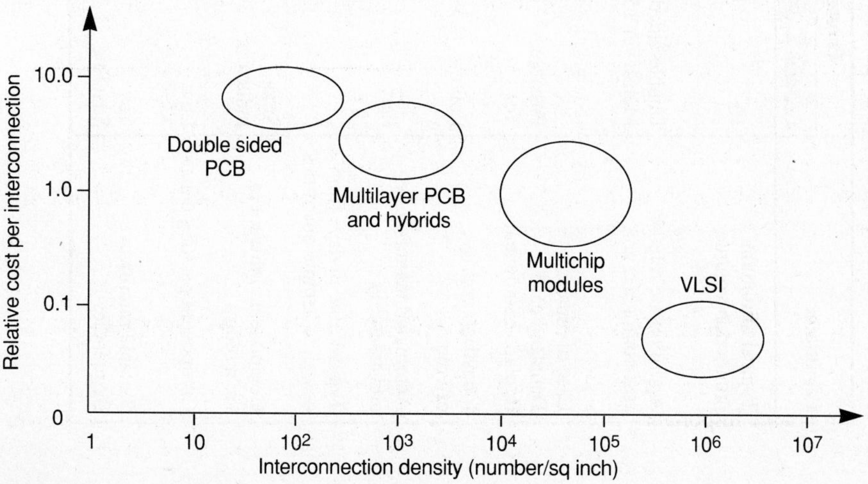

Figure 1.12 Comparison of manufacturing costs

Table 1.1 Comparison of manufacturing technologies

Parameter	Printed circuit board		Wire wrap	Multiwire	Film hybrids		ASIC
	Leaded	SMT			Thick	Thin	
Typical minimum production volume	10	10 Manual 100 Automated	1 Manual 100 Automated	≈ 5	10	100	1000
Approximate relative cost completed unit	Double-sided 1 Multilayer 15	Single-sided ¾ Multilayer 10	10	4	1	5	Small volume 2 Large volume ¼
Ease of repair: Damaged card Replace component	 Possible Yes	 Throw away Yes	 Yes Easy	 Possible Yes	 Possible Possible	 Throw away Possible	 Throw away
Industrial security offered	Low	Low	Low	Low	Medium	Medium	High
Automated manufacture and assembly	Part	Full	Part	Part	Full	No	No
Typical time in days from design to sample quantities	7	7	2	7	14	14	7 → 70
Components inherent to technology	Small inductors	Small inductors	—	—	Resistors, small capacitors and inductors	Resistors, small capacitors and inductors	Transistors, diodes and resistors
Component packing density	Low	Medium	Low	Low	Medium	Medium	High
Suitable microwave frequencies	Yes Stripline	Yes Stripline	No	No	Yes Microstrip	Yes Microstrip	Yes Si ECL and GaAs

This book will concentrate on printed circuit boards and related processes. The topic of application-specific integrated circuits and film hybrids are covered in other texts (Haskard 1990; 1988), while the multiwire system is a special commercial process that can be considered as a low-cost multilayer board where only low volumes are needed.

1.6 Questions

1. For what reasons did printed circuit board technology supersede direct wiring using tag strips? Why do printed circuit boards achieve repeatable electrical performance?
2. As integrated circuit package pin counts increase, the length of interconnecting wire also increases, resulting in the need for multilayer interconnection. Compare three of these processes: multilayer printed circuit boards, wire-wrap and multiwire systems.
3. Why have surface mount printed circuit boards proved to be more difficult to manufacture than at first anticipated?
4. Explain the 'mother-daughter' board construction concept. How does it differ from the printed circuit back plane assembly method? Are the two processes complementary?
5. Integrated circuit speeds are increasing and digital systems today can have clock frequencies in the VHF range. What impact do you believe this will have on printed circuit board technology?
6. List the advantages and disadvantages of employing a silicon wafer instead of a printed circuit board to interconnect silicon dies to form a system.
7. As the design engineer for the electronic control portion of company products, you need to make recommendations on the manufacturing method to be used. Three products are to be considered: (a) a toy train; (b) a washing machine; and (c) a radar system. Figure 1.9 lists the various manufacturing alternatives available. What are the most important considerations for each of the three products and your recommendations?

1.7 References

Coutts, T. J. (1978), *Active and Passive Thin Film Devices*, Academic Press, London.

Dettmer, R. (1988), 'The silicon PCB', *IEE Review,* **34**(10), pp. 411–13.

Geschwind, G. & Clary, R. M. (1991), 'A comparison of MCM–D and PCB costs', *Inside ISHM,* **18**(3), pp. 26–31.

Haskard, M. R. (1988), *Thick Film Hybrids: Manufacture and Design*, Prentice Hall, New York.

Haskard, M. R. (1990), *An Introduction to Application Specific Integrated Circuits*, Prentice Hall, New York.

Jesshope, C. & Moore, W. R. (eds), (1986), *Wafer Scale Integration*, Adam Hilger, Bristol.

Maissel, L. & Glang R. (eds), (1970), *Handbook of Thin Film Technology*, McGraw-Hill, New York.

Seraphin, D.P. (1978), 'Chip module package interfaces', *IEEE Trans on Components: Hybrids and Manufacturing Technology*, CHMT **1**(3), pp. 305–9.

2 Printed circuit board technology

2.1 Introduction

The card most commonly used in the electronics industry to assemble components is the printed circuit board (or printed wiring board). It is a synthetic, laminated, insulating material to which copper tracks have been added. While there are various types of boards, the simplest is the single-sided board. The steps to manufacture it are illustrated in Figure 2.1.

Initially there are two separate paths, the design of the board layout and the bare board manufacture. While the design of the layout will be discussed in Chapter 10, it is important to realize that, in addition to the actual track layout, the designer must supply a considerable amount of other information. The design documentation must include the following minimum information:

1. Board identification number/title.
2. A statement on the board type—single layer, double-sided plated through hole, multilayer, etc.
3. Specification for the bare board material to be used, insulation and conductor material, quality and thickness.
4. Track details for each conductor layer. If there are several layers, alignment marks must be included. The scale is to be given.
5. Drill sizes and hole positions.
6. Board finish: final dimensions, solder mask, screened component details, etc.

Documentation can be generated manually or by using computer aided design methods. The medium of transfer of this information can therefore vary considerably. Take the track layout information: it can be provided on self-adhesive paper crepe material, stuck to a stable based plastic sheet such as polyester, a computer plot or an electronic transfer using magnetic tape, floppy disc or cartridge. Frequently the artwork is prepared on an enlarged scale. Manual drawings are usually enlarged two times, but fine linework may be enlarged four or even 10 times to improve accuracy. Even with computer generated artwork these scales may be used because final accuracy depends on plotter positioning accuracy, repeatability, and pen or aperture (for a photo plotter) sizes. In all such cases, the artwork is reduced to the final size in one or more photographic step, depending on whether a positive or negative is required for the board manufacture. Appendices A and B show artwork for single-sided boards and Figure 2.2 shows some typical artwork for a double-sided plated through-hole board.

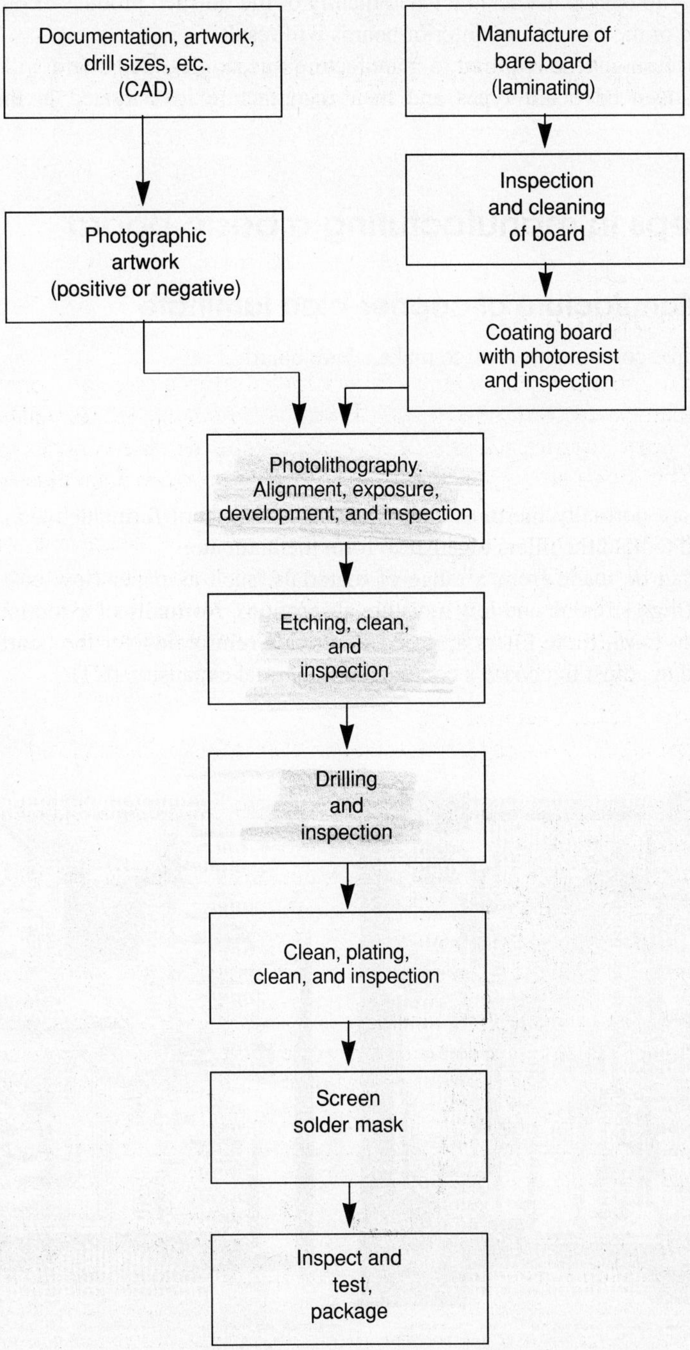

Figure 2.1 Steps in the manufacture of a single-sided printed circuit board

The processing shown in Figure 2.1 has several cleaning and inspection steps. These are necessary to ensure the accuracy and quality of the finished product. If omitted or left until the end of the processing, inferior boards will result.

The processing steps required to manufacture this single-sided board will be outlined and more advanced board types and their manufacture investigated in the following discussion.

2.2 Steps in manufacturing a basic board

2.2.1 Manufacture of copper-clad laminate

There are three components used to make a bare board:

1. resin
2. filler
3. copper foil.

The resins are normally thermosetting types such as phenol formaldehyde or an epoxy and are used to hold the fillers together to form the laminate.

Fillers can be made from a range of materials, such as paper (low cost), cotton or glass cloth (high strength and low moisture absorption). Normally of a continuous web or woven fabric form, these fillers are used to provide reinforcing for the board. Additives may be used to adjust the board's coefficient of thermal expansion (CTE).

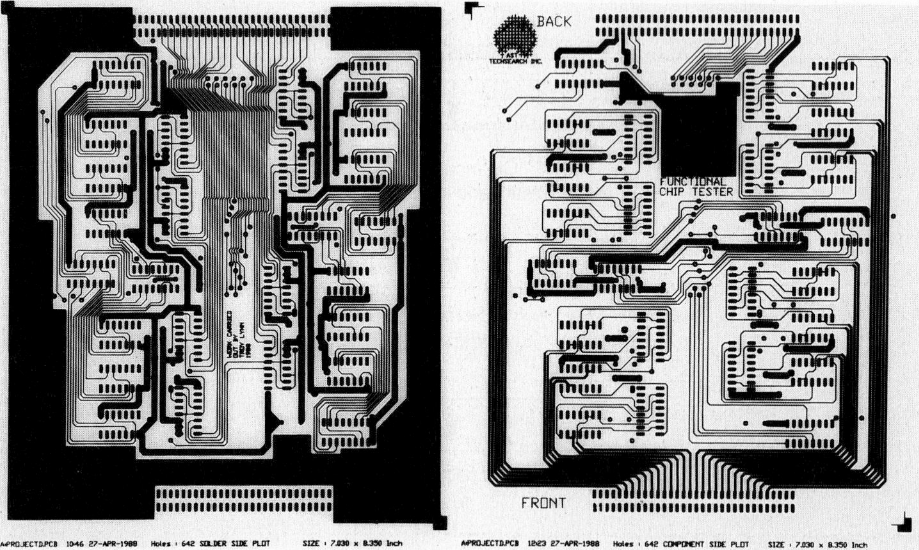

Figure 2.2 Manually generated artwork using Bishop's tape on a Mylar backing for a double-sided plated through-hole board

Table 2.1 Selected properties of various laminated boards

Properties	Phenolic paper base	Epoxy glass base	PTFE glass base	Polyester paper base	Polyester glass base	Silicone glass base	Unit
Insulation resistance	3.10^7	10^6			10^5	10^6	M ohms
Dielectric strength	10	15	15	20	15	10	kV/mm
Dielectric constant 1 M Hz	4.5	5	2.5	4	5	4	—
Dissipation factor 1 M Hz	0.05	0.02		0.025	0.01	0.075	—
Tensile strength	1000	3000	1000	800	4000	1700	kp/cm^2
Flexural strength	1500	4000	900	1000	4500	1700	kp/cm^2
Shear strength	800	1200	750		2000	1200	kp/cm^2
Water absorption	0.4	0.1	0.02	2	0.8	0.3	%
Specific gravity	1.3	1.8	2.2	1.3	1.9	1.75	kg/dm^3
Machining qualities	Good	Good		Excellent	Fair	Fair	
Thermal conductiv.	4.5	7.2	1.5		7.1	710	10^4cal/sec cm^2/°C/cm
Specific heat	0.4	0.375	0.2		0.3	0.25	cal/°C/g
Thermal expansion		1		3	2	0.75	$°C.10^5$
Effect weak acids	None	None	None	None	None	None	—
Effect strong acids	Slight	Slight	Attacked	Some attack	Some attack	Some attack	—
Effect weak alkalines	Slight	Slight	None	Slight	Slight	Slight	—
Effect strong alkalines	Attacked	Attacked	Attacked	Attacked	Attacked	Attacked	—
Effect organic solvents	None	Slight	None	None	Slight	Attacked	—

The copper foil is normally specified by weight, that is, ½ ounce per square foot (152.5 g/m^2), one ounce per square foot (305 g/m^2) or two ounces per square foot

Table 2.2 Maximum temperature of various laminated boards

Resin	Filler	Maximum temperature °C
Phenolic	Paper	100
	Glass	250
Epoxy	Glass	120
Polyester	Glass	120
Silicone	Glass	250
Teflon	Glass	200

(610 g/m^2). These correspond to foil thicknesses of 17.5, 35 and 70 microns respectively. It is usually produced by electrolytic deposition on a flat mandril, and is about 99.8% pure and has a tolerance on thickness of about ± 10%.

The laminating of these materials to form the board is as follows. The filler in web or cloth form is first impregnated with the thermosetting resin by dipping it into a solution of the resin and then squeeze rolling. The laminate is partially cured by heating to drive off solvents so that it is tack-free. These laminates are then cut into sheets. Next they are stacked between copper sheets (for double-sided boards), the number of laminate sheets and thickness of the copper sheet determining the final board thickness and copper weight. The sheets are placed between steel plates in a press where they are heated to 120–170°C under a pressure of 20–110 kg/cm^2. The resin flows and cures permanently. After cooling the sheets are trimmed, inspected for adequate quality, cut into smaller sheets about 3 feet square and vacuum sandwiched between clear plastic sheeting. This provides protection for the copper surface, particularly from oxidization by air. Boards are rejected at inspection stage if they have warp and twist, imperfection in the copper surface, or poor bonding between the copper and the laminate.

The final properties of the laminate depend upon the materials used and process control during manufacture. In addition to electrical properties such as dielectric strength and constant, dissipation factor, insulation resistance, resistivity (both surface and volume), there are physical characteristics. These include flexural strength, punching and drilling qualities, flame resistance, and water absorption. Table 2.1 shows selected properties for a range of laminates. The figures given are typical mid-range values (Harper 1970; Bosshart 1983).

Various standards define how tests are to be performed on copper-clad laminates. These include the AES, BES4584, MILL-P-13949D and IEC249.

One important non-electrical characteristic is the maximum temperature at which boards can be operated. This is also relevant for the curing of printed epoxies and thick film pastes on the boards. Table 2.2 gives some temperatures.

2.2.2 Cleaning of boards

Before any processing can be undertaken on a board, it must be cleaned. There are three main areas of contaminants:

1. Organic material: these are mainly oils and greases arising from equipment used in mechanically 'shaping' the board (punching, drilling) and human oils (sweat marks, finger prints). Degreasing solvents must be used to remove this organic material.

2. Oxides: clean copper surface readily oxidizes in air, resulting in a surface that prevents chemical processing. In an industrial environment other corrosive gases and compounds may be present in the air, so that materials like sulfides may also be formed. Thick oxides and sulfides can only be removed by using abrasives and scrubbing, while chemical etching may be used to remove thin layers.
3. Particulate matter: dust, machining particles and other small particulate matter often adheres to a board surface. It can be removed by washing in water, blowing with compressed air or brushing.

A cleaning process must therefore be established which allows for all three types of contaminants.

For small-scale production the cleaning may be undertaken manually, but for moderate and large production levels special cleaning machines are needed. Figure 2.3 shows typical cleaning steps.

The cleaning process begins with a wash in a solvent or degreasing solvent solution to remove all grease and oils. This may be a spray or dip process. After a wash in water (normally tap water) the laminated board goes into a scrubbing stage to remove any oxides or sulfides on the surface. Pumice slurry or abrasive cleaners similar to the domestic types are used. Care must be taken not to scratch the surface because scratches deeper than ½ micron or so can degrade the performance and appearance of the final board. After washing in water again (normally tap water) a softer mechanical process of wet brushing may be used. Next follows an acid dip in hydrochloric acid to remove residual alkali and metal oxides to prepare the surface for good resist adhesion.

The board now goes through a series of washes with high quality deionized water of at least 100 kΩ/square resistivity (across opposite faces of a meter cube). Finally, the board is dried in an oven. Figure 2.4 shows a commercial board cleaning plant.

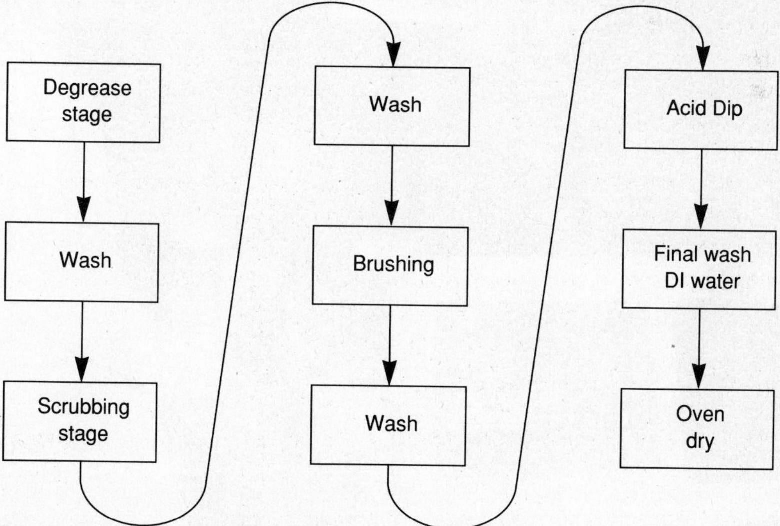

Figure 2.3 Typical cleaning steps for an unprocessed laminate board

2.2.3 Photolithography

In this process the original artwork pattern is transferred to the copper on the card. As already mentioned, the artwork must be in the form of a photographic negative or positive. The phototooling film, as it is called, must have:

1. Dimensional stability so that its size does not change appreciably with temperature and humidity.
2. Line edge quality, i.e., achieving a high degree of contrast at the transition between clear and dark areas on the film.
3. Good resolving power, with 2,500–10,000 lines/inch (100–400 lines/mm) usual.
4. High exposure latitude, i.e., tolerant on changes to exposure times.
5. Durability, so that the film can withstand the necessary process handling.

Most films consist of a transparent backing of polyester, typically 175 micron (7 mil) thick with a light sensitive silver halide emulsion 4–8 microns thick (Spitz 1987). The wavelength of maximum sensitivity is about 480–550 ångströms. The safe light is in the red wavelength region, so processing of the film is often undertaken in what is called a red room.

The cleaned copper board is coated with a photoresist which is sensitive to ultraviolet light (2000–5000 Å). Again, like the phototooling film, the photoresist has a safe visible light area in the yellow to red wavelength (5600–7000 Å).

There are two basic types of photoresists: wet and dry. Wet resists are applied as a solution and the resultant coating thickness is set by the viscosity of the resist and the method of application. The resist may be applied by one of the following techniques:

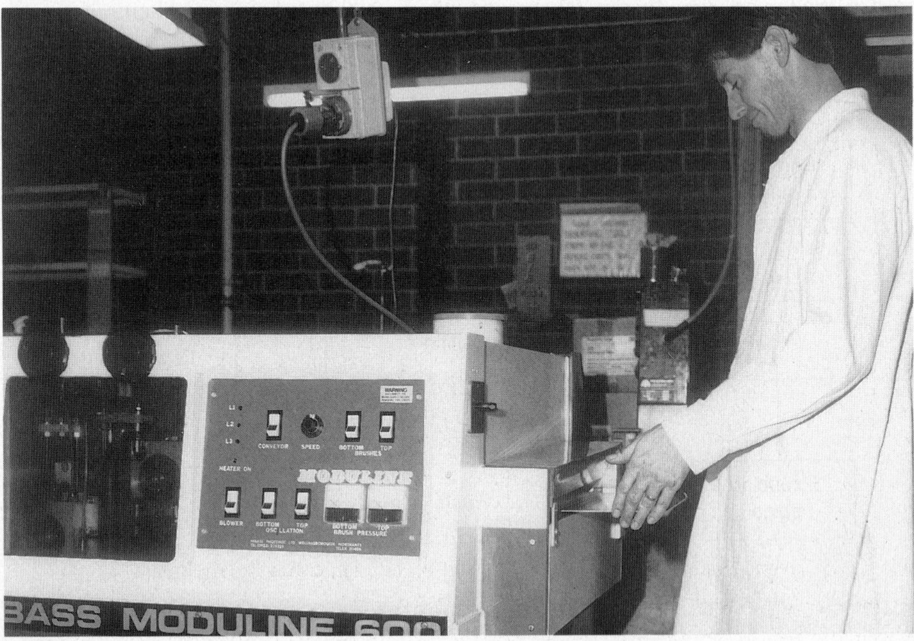

Figure 2.4 Cleaning of laminate boards ready for processing
(Reproduced with permission of IMP Engineering)

1. Flow: the resist is placed on the board and it is tilted until the resist covers the whole surface. Control of the thickness is poor.
2. Dip: the board is dipped into the resist and pulled slowly out vertically. The coating is usually thicker at the bottom of the board than the top.
3. Roll: resist on the roller coats the board. The method is suitable for large area boards.
4. Spin on or whirl: the board is spun at about 150 rpm and the resist on the center of the board spreads out through the centrifugal force. Excess resist is spun off and wasted. The film is thickest in the center of the board and thinnest at the outer edges.
5. Spray on: multiple spraying can build up the coating thickness. Drilled holes in the board do not get clogged with resist.

After applying the resist it is normally baked at 80°C for a few minutes to harden.

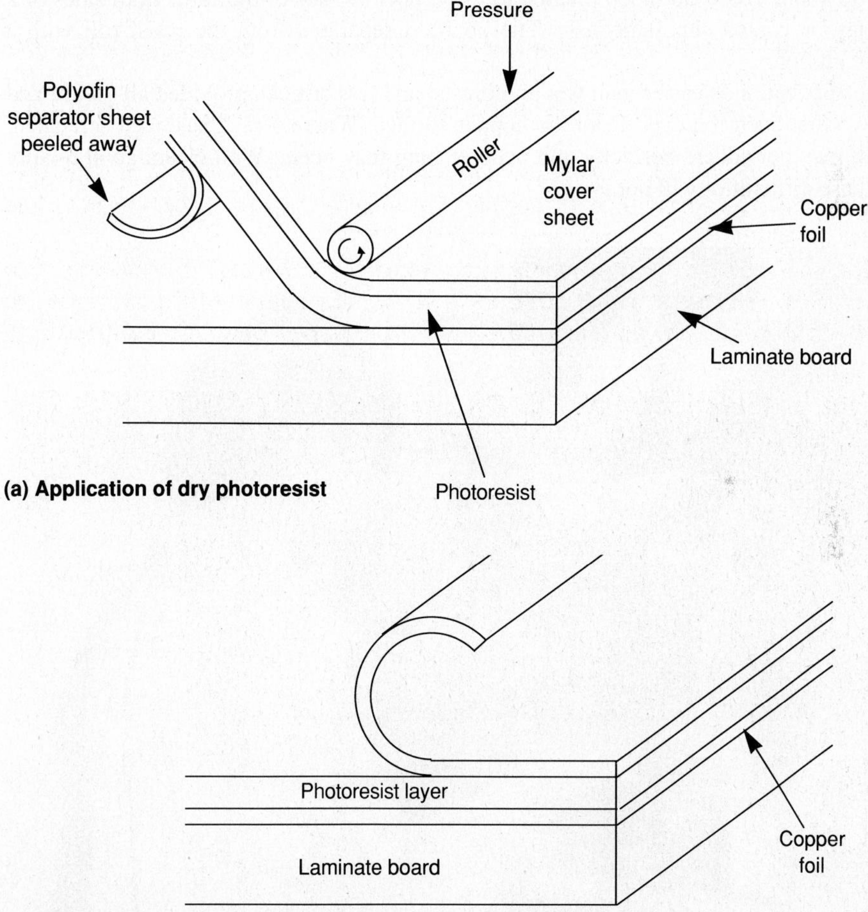

Figure 2.5 Application and processing of a dry film resist. (a) Application of dry photoresist. (b) Removal of cover sheet. Coated board now ready for exposure and development

The resist may be of negative type (such as KPR, KMER) where the ultraviolet light polymerizes the resist so that it becomes insoluble in the developer. The artwork must therefore be in the form of a negative. Positive resists (originally made by Shipley as a series of AZ types) can also be employed. Here the polymerized resist is soluble in the developer and the artwork must be in the form of a positive.

Late in the 1960s a new form of photoresist was introduced by Du Pont called dry film resist. It is a three-layer sandwich, the center layer being the photoresist (available in thicknesses from 12–70 microns), an upper polyester cover film, and a lower polyolefine separator layer which is removed just prior to the film's application to the card. This is illustrated in Figure 2.5.

The application of the dry film resist is carried out in a machine called a laminator (see Figure 2.6). The separator is automatically stripped off and the photoresist is heated briefly to about 110°C and then pressed to the copper surface of the card. Both sides of a card may be coated simultaneously. The board is separated from the resist roll with a knife or guillotine.

The application is easier than wet photoresist and less critical, provided all grease and oils have first been removed from the copper surface. Where a card has been scratched, the film may not adhere perfectly and underetching may occur. With clean, good quality cards these difficulties will not arise.

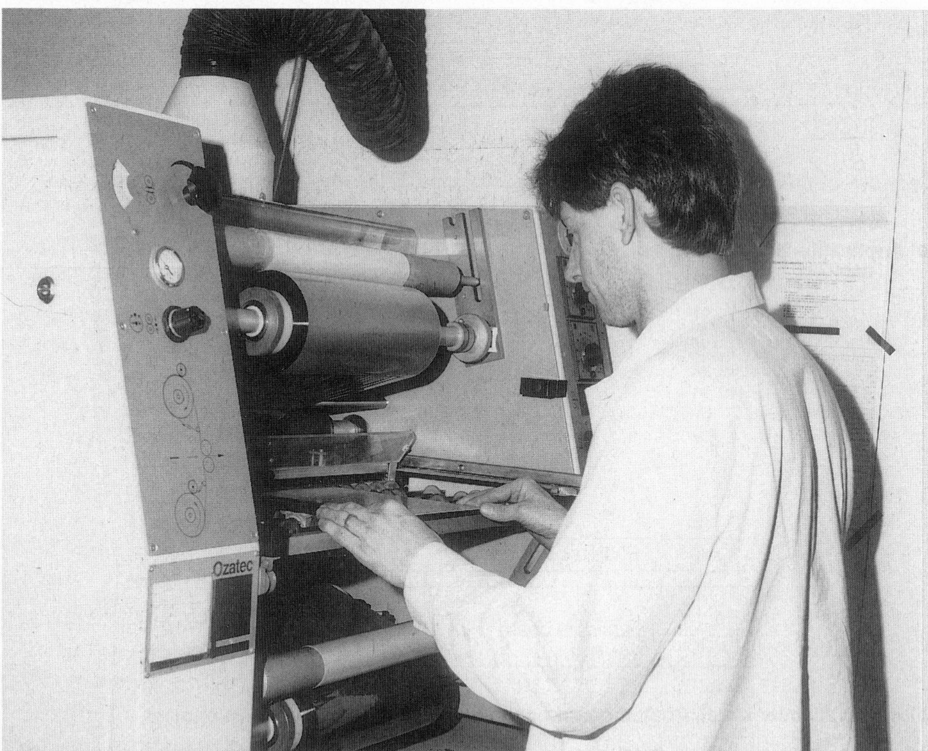

Figure 2.6 Laminating machine used to apply dry film resist to a board (Reproduced with permission of IMP Engineering)

Table 2.3 Comparison of liquid resist and dry film photolithography

Process	Liquid resist	Dry film
Preparation	√	√
Application	Coat	Laminate
Oven dry	√	
Expose	√	√
Develop	√	√
Touch up	√	√
Dye	√	
Wash and dry	√	
Inspect	√	√
Post-bake	√	
Etch	√	√
Strip resist	√	√

There are two types of dry film resists: those developed in a solvent such as methylchloroform and those developed in aqueous solutions. The latter type is gaining in popularity as there are no solvents to dispose of (Smith-Vargo 1987). Both dry film resists are of a negative type. It is not necessary to bake the resist because it is applied with heat.

Regardless of whether wet or dry resists are used, the following steps apply. The coated board is exposed through the appropriate negative or positive to ultraviolet light. If the board is double-sided, a double-sided exposure is made. Care must be taken to ensure that the patterns on both sides are aligned and this is normally achieved by matching up special punch holes on the boards and photographic films. Exposure times can range from a few tens of seconds to several minutes. The resist is then developed, leaving those portions of the copper to be retained on the board covered by the resist. Table 2.3 compares image transfer processes.

The liquid resist process has two additional steps: a dye and a post-bake stage. The dye is optional and is used to dye the resist so that it is easy to see on the board. The post-bake, usually at about 100°C for 10 minutes, hardens the resist before the etching process. Stripping after etching is normally a scrub with a solvent or commercial stripping solution, followed by a rinse in deionized water. Touch up is the process of correcting any minor defects. Protective coatings are hand painted on areas where the resist has been removed. Too much touch up means that the photo process needs close examination.

2.2.4 Etching of the board

In the etching process the unprotected copper foil is chemically removed. Unfortunately, the very etching process exposes the sides of the copper so that the copper is etched under the resist protective layers. The time taken for the etch is therefore critical. Too little can mean not all the copper has been removed, so shorts between tracks remain; too long can mean that track widths are reduced through overetching. If this occurs, tracks may be unable to carry the desired current, present an incorrect impedance (stripline work) or, in the worst case, become open circuit (see Figure 2.7).

Table 2.4 Comparison of board etchants (Harper 1970)

Etchant	Corrosive-ness	Neutralisation disposal problems	Toxicity	Ventilation requirements	Post-etch cleaning difficulties
Ferric chloride	High	Medium	Low	Low	Medium
Ammonium persulfate	Low	Low	Low	Low	Low
Chromic acid	High	High	High	Medium	Medium
Cupric chloride	High	Low	Medium	High	Medium
Alkaline ammonia	High	Medium	Medium	High	Low

There are several chemicals used for etching. The most common are shown in Table 2.4. The oldest and perhaps still the most common etchant is ferric chloride. It normally comes in crystal form and the crystals are dissolved in deionized water to achieve the desired concentration. This is typically 500 g of ferric chloride per liter of water. Unfortunately, the material is corrosive and leaves dark stains. Further, it is difficult to regenerate for reuse. If solder or tin is used as a mask ferric chloride will attack it.

The chemistry of the process is as follows:

1. The copper is converted to cuprous chloride.

$$FeCl_3 + Cu \rightarrow FeCl_2 + CuCl$$

2. The cuprous chloride is further converted in the etchant solution to cupric chloride.

$$FeCl_3 + CuCl \rightarrow FeCl_2 + CuCl_2$$

3. This cupric chloride also reacts with the copper to form cuprous chloride.

$$CuCl_2 + Cu \rightarrow 2\ CuCl$$

4. Free acid is usually present through the hydrolysis reaction.

$$FeCl_3 + 3\ H_2O \rightarrow Fe(OH)_3 + 3\ HCl$$

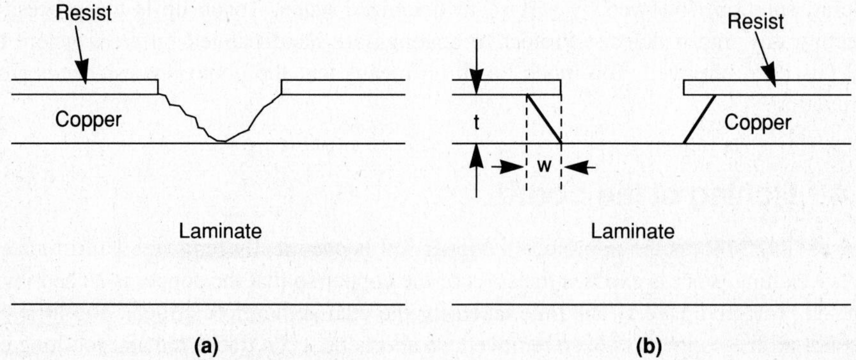

Figure 2.7 Etching of a printed circuit board illustrating: (a) underetching, and (b) undercutting through over etching. The ratio w/t is called the etching factor

The level of hydrochloric acid can be increased to hold back the formation of insoluble ferric hydroxide and thereby improve the etching time. The time taken for an etch largely depends on the concentration of the solution and temperature.

Ammonium persulfate is frequently used because it has few of the disadvantages of ferric chloride. Solder and tin can be used as masks. The overall reaction that takes place when etching copper is:

$$(NH_4)_2\,S_2O_8 + Cu \rightarrow Cu\,SO_4 + (NH_4)_2\,SO_4$$

Unfortunately, while it is a strong oxidizing agent, ammonium persulfate is unstable in solution and decomposes to form hydrogen peroxide, oxygen and peroxydisulfuric acid. The latter is a slow oxidizer at room temperature and catalysts involving transition metals such as manganese, iron, and mercury are added to assist. The reaction rate is a function of the particular catalyst and the solution pH. Solutions are typically made up of 240 g of ammonium persulfate dissolved in every liter of deionized water. Catalysts such as mercuric chloride may be added at 27g/liter (Coombs 1967). The solution tends to be unstable, can produce salt crystals on boards and, if there is a high lead content solder on the board, will cause a white discoloration on the solder surface.

The etchants may be applied to boards in one of several ways—immersion, bubble, splash and spray, the latter being by far the most common method. The etcher consists of a sealed spray box in which the boards are hung vertically so both sides can be etched simultaneously. More even etching is achieved when either the spray nozzles or the board are slowly oscillated or rotated. The pipes and materials used to construct the unit must

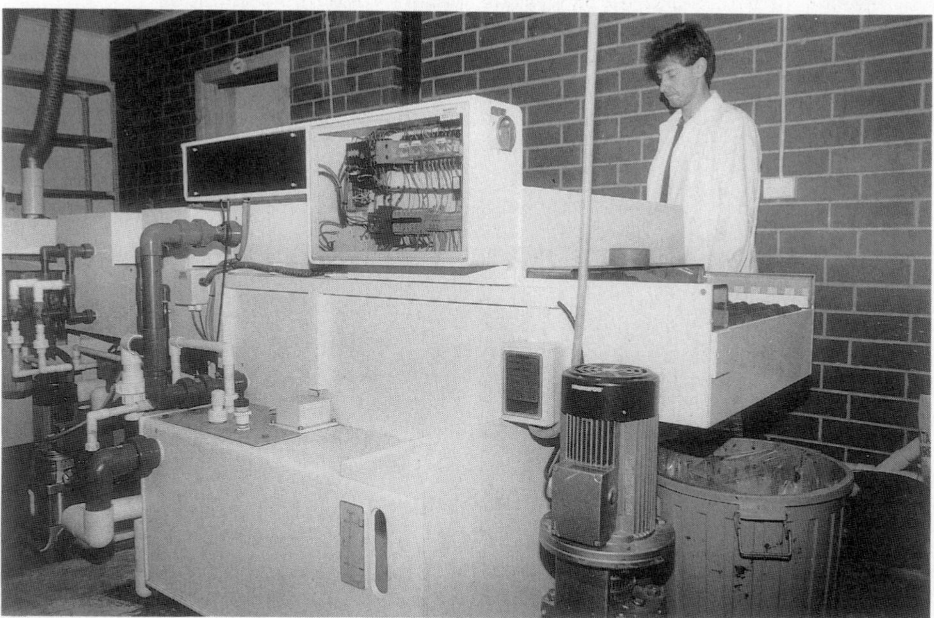

Figure 2.8 Printed circuit board spray etcher
(Reproduced with permission of IMP Engineering)

be such that they are resistant to the corrosive etchants. Polypropylene and PVC are commonly used. On completion of the etch, boards must be washed and inspected. Figure 2.8 shows a commercial etching unit.

2.2.5 Board drilling

Boards can be drilled manually with or without the use of jigs to ensure correct drill sizes are used and that no holes are missed. Cards are normally stacked so that several are drilled simultaneously. Tungsten carbide or diamond tipped drills are preferred for fiberglass boards. The vias and pads normally have copper etched from the center to help center the drill. Some manual machines drill from the underside and use a light spot on top to indicate the drilling point.

For large-scale production, punching or drilling is used. With punching, the correct laminate must be used: a phenolic or polycarbonate. The minimum size hole that can be punched in a $\frac{1}{16}$ inch (1.6 mm) thick board is $\frac{1}{32}$ inch (0.8 mm), so that leaded components must be clinched before soldering. The cards may also be punched on the edges with a row of holes on one or more sides to allow the card to be broken into individual biscuits (Figure 2.9).

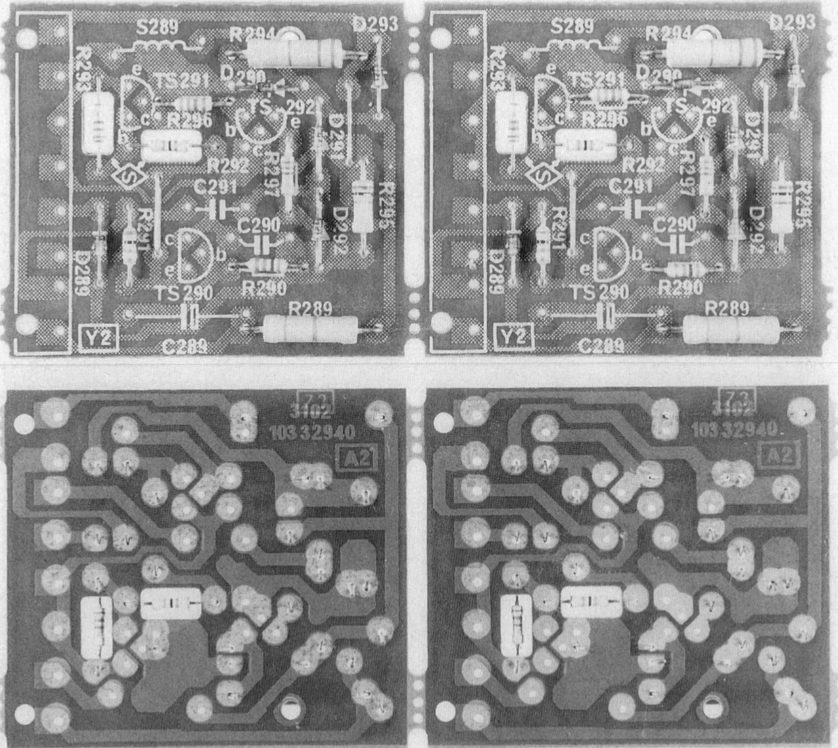

Figure 2.9 An array of cards that have been punched. The closely spaced holes at the edges allow the boards to be broken into individual cards

Drilling of production quantities is normally done with a numerically controlled multihead drilling machine. Machines with two or four heads, each simultaneously drilling a stack of cards, are typical. Dummy cards are placed at the top and bottom

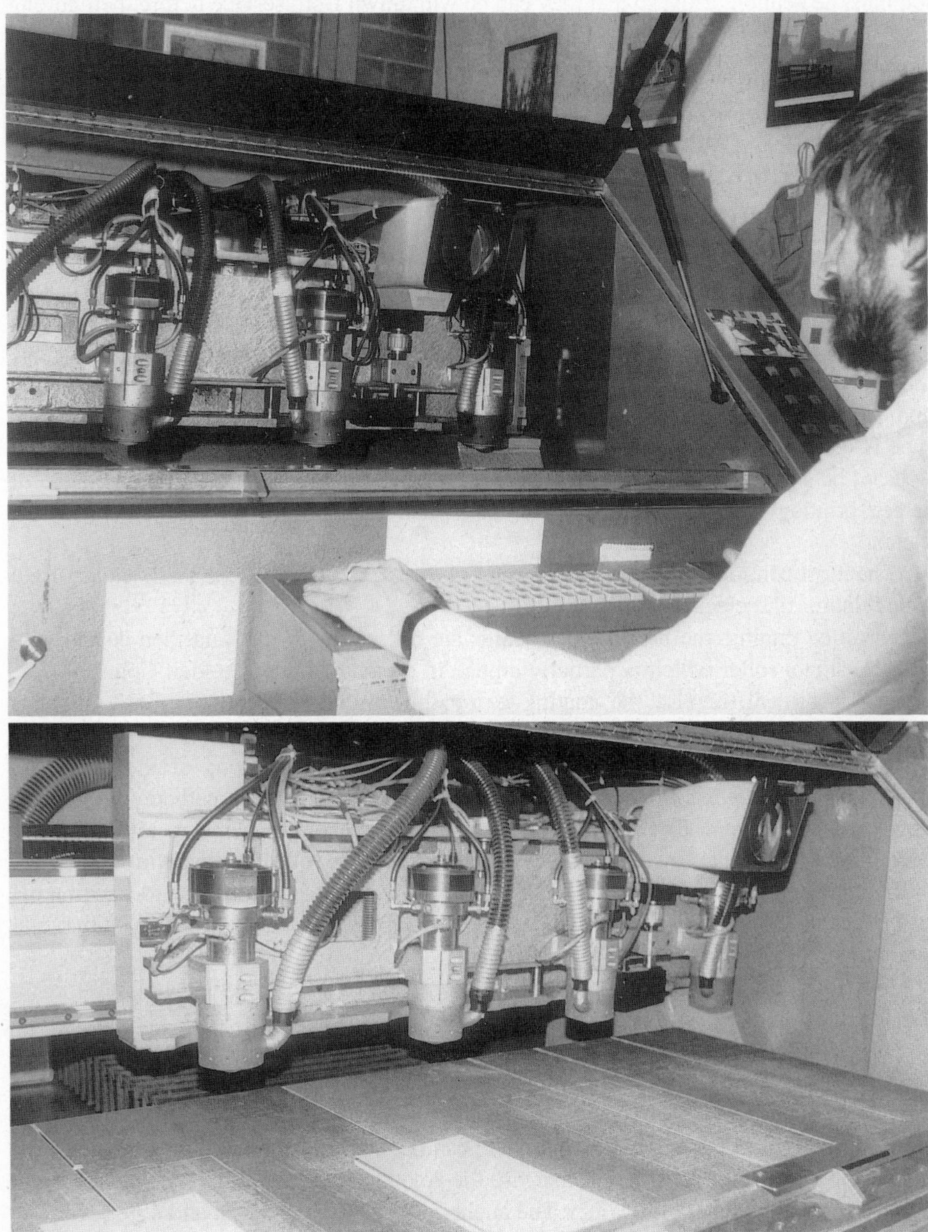

Figure 2.10 A numerically controlled drilling machine
(Reproduced with permission of IMP Engineering)

of each stack to ensure consistency of drilling. Tapes for the drills are prepared directly from the computer aided design software, or in the case of manual artwork, using a digitizer. The numerically controlled drill automatically changes drills, adjusts drill speed and feed rate, and positions the card stacks to the correct X–Y coordinates for drilling. A log must be kept on drills, so that they can be withdrawn from service before their cutting edge deteriorates and causes burrs. Small drills are discarded and the larger, more expensive ones resharpened. Figure 2.10 shows a numerically controlled drilling machine.

2.2.6 Coatings

While bare copper boards are acceptable for experimental and prototyping work, they are not normally acceptable for high quality products. The copper, even when it has a protective organic coating (such as the photoresist), will deteriorate with time, making soldering more difficult and having an inferior appearance. Coatings are normally added to prevent these problems. They may be categorized into two types: metal and organic materials. Metals include gold plating for connectors, tinning of the tracks, nickel plating and, in some instances, copper plating to build up some of the track areas. The organic type is usually a solder mask, a screen-printed epoxy coating to cover those areas that will not be soldered. It may follow or precede the coating with metals. The principle of screen printing is discussed in Chapter 6. In this section we will concentrate on the metals.

The most common coating is the tinning of the copper tracks. Two methods are used: rolled tinning and electroplating.

In rolled tinning, the bare copper boards are brushed with flux and then coated with solder, using a roller which is partially dipped in a bath of molten solder. Control of the process is critical, or else the coating is too thin or the temperature of the board is excessive and fine line tracks peel off.

Electroplating is carried out in an electrolytic cell where ions of the metal are transferred from the electrolyte and deposited onto the copper track cathode through the flow of a DC current. The anodes are frequently made of the same metal so that no contaminates are introduced. Stainless steel hooks are used to support and make connections to the electrodes. For some baths the electrolyte should be acidic with a pH $\geq$ 8. Exceptions are the gold and nickel electroplating. The current density, time, and temperature determine the thickness and quality of the plating.

For tin plating, the electrolyte contains a stannous sulfate solution and, at ambient temperatures and current densities of 20Å/square ft (200Å/m^2), the plating rate is typically 1 micron per minute.

Should the plating required be solder—that is, a tin/lead mixture—the electrolyte is a mixture of stannous fluoroborate and lead fluoroborate. Peptone is added to give a fine grain finish. To achieve a near 60/40 mix deposition, control of the plating unit is critical. Increasing the current density will increase the amount of tin deposited. A typical current density is 15Å/square ft (150Å/m^2). The anode is an alloy of 60% tin and 40% lead. The tin/lead mixture resulting from plating tends to be porous. This can be overcome by passing the boards through a hot air leveling unit (see Figure 2.11). The boards are dipped vertically into a solder bath and, as they are withdrawn, subjected to hot air blasts. This removes all excess solder and clears vias and holes, leaving a high quality flat solder surface.

When gold plating is required, it is usual to first give the board a nickel plating. The electrolyte contains nickel sulfate and nickel chloride, having an alkaline pH of about 3.5. The temperature is usually elevated to 45–60°C and current density of the order of 40–60 Å/square ft (400–600 Å/m^2).

Being a soft metal, gold requires hardening additives so that it will have a useful lifetime on connector pins. Hardeners of indium, cobalt, or nickel are used. The electrolyte is therefore mainly a solution of potassium alluvial cyanide with a hardener. The pH will be typically 4, bath temperature 30–40°C and current density 10 Å/square ft (100 Å/m^2), resulting in deposition rates of 0.4 micron/minute. If a straight gold layer is used over the copper track, its thickness is typically 5 microns. Should a nickel undercoat of 4 microns be used, the gold thickness can be reduced to 2 microns.

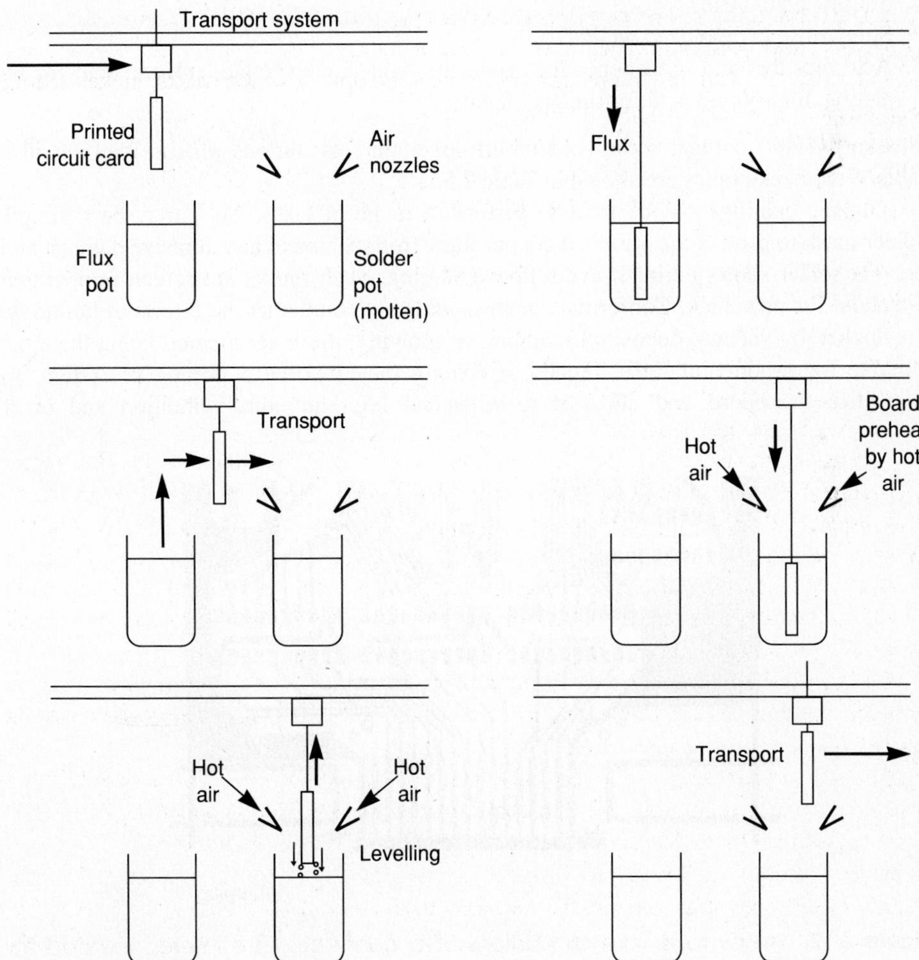

Figure 2.11 Hot air leveling system

Before any plating is undertaken there are two important factors that need to be considered. First, the board portions to be plated must be electrically connected for a current to flow in them. For example, a connector to be gold plated may be secured with a strap to ensure all portions are interconnected. The strap is later removed by guillotining. See the example in Figure 2.12. Second, not only is the board to be washed clean, but the copper surface etched to ensure there is a new, clean surface. Bosshart (1983) describes a typical preplate clean:

1. Degrease with organic solvent such as a 60/40 mix of trichloroethylene and perchloroethylene.
2. Alkali clean in a sodium hydroxide bath.
3. Spray rinse with deionized water.
4. Ammonium persulfate copper etch.
5. Spray rinse in deionized water.
6. Sulfuric acid clean.
7. Spray rinse in deionized water.
8. Acid dip, the acid used depending upon the bath. For example, sulfuric acid for tin plating, fluoroboric acid for tin lead plating, etc.

Bosshart (1983) compares the solderability over time of various plating methods in a table. Several examples are shown in Table 2.5.

Organic coatings can be used to perform a range of tasks. They may be a simple solder mask to protect the untinned copper track from chemical and abrasive damage and prevent solder shorts during wave reflow soldering. Such masks are screen printed and are about 0.1 mm thick. Conformal coatings, added to counteract the effects of humidity, are applied by vacuum deposition, dipping or spraying, the latter method being the most suitable for production lines. Finally, a coating may involve a potting procedure, to ruggedize the board and allow it to withstand high humidity, vibration and other

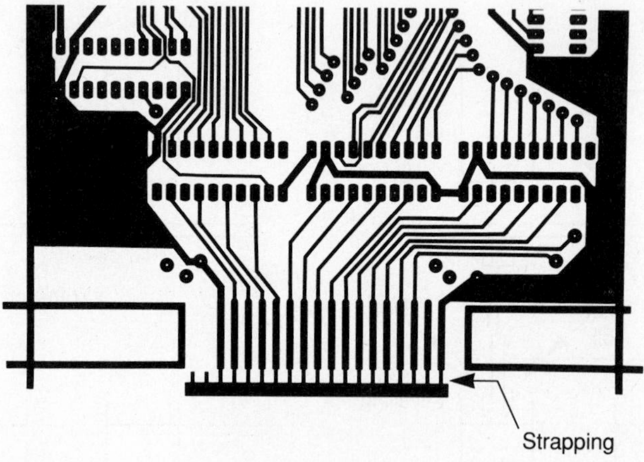

Figure 2.12 Straps to ensure all portions of a connector are interconnected for gold plating to occur over the whole connector. The strap is removed when the board is guillotined to size

Table 2.5 Comparisons of various coatings on a board. Solderability index is defined as 2/(wetting time in seconds)

Coating	Solderability index			
	Freshly deposited	**After 24 hours in steam**	**21 days 95% RH 40°C**	**6 months normal storage**
Tin, 8μm	10+	10+	9.5	9
60/40 Tin lead, 8μm	10+	10+	10	10
60/40 Roller coat	10+	10+	3	3
Hard gold, 5μm	10+	5.5	10	4.5
Lacquered copper	10+	6.5	Non-wetting	4

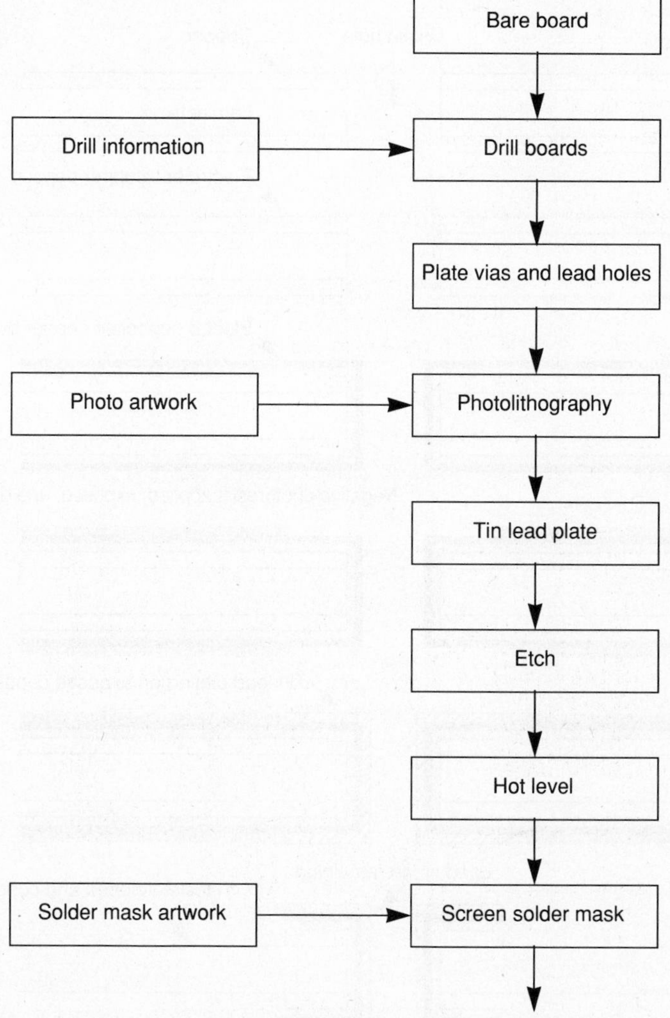

Figure 2.13 Major steps in the production of a double-sided, plated through-hole board

mechanical stresses for long periods. All the materials used are based on phenolics, acrylics, epoxies, polyimides, polyurethanes and silicones, with epoxy and polyurethane-based coatings being the most common (Coombs 1967; Lea 1988). In cases which require the highest quality conformal coatings, vacuum deposited Paraxylylene is used because it penetrates all cracks and crevices, gives a thin, even coat, and provides excellent chemical and abrasive protection.

2.3 Other printed board processes

The processing steps described earlier apply to most board processing. Depending on the board type, the order may be changed and additional steps introduced. Other board types that are in common use today are:

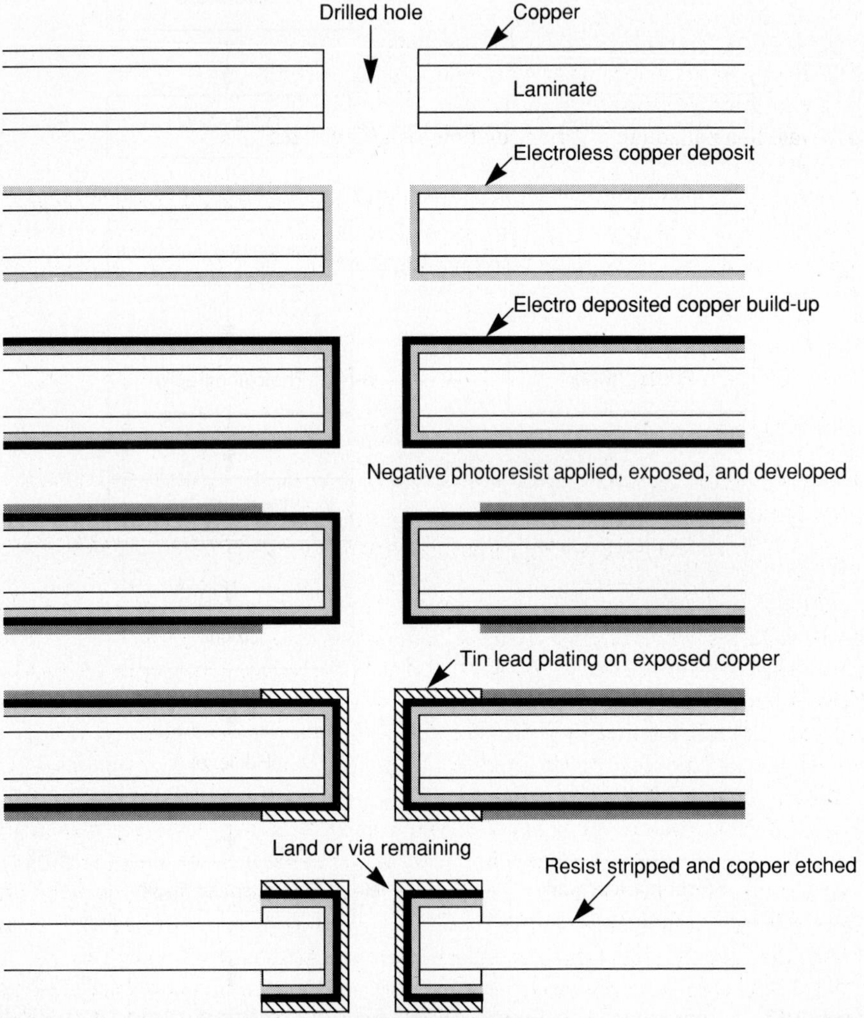

Figure 2.14 Plated through-hole processing sequence

1. double-sided, plated through-hole boards
2. multilayer boards
3. flexible boards

The flow diagram for the double-sided, plated through-hole board is shown in Figure 2.13 (Karpel 1988). For simplicity, cleaning and inspection stages have not been included in this diagram.

The additional steps occur at the beginning, where boards are drilled and plated to form the plated holes and vias. Figure 2.14 shows these steps.

The double-sided blank board is first drilled and then any burrs removed. The board is thoroughly cleaned and prepared for plating (see Section 2.2.6). Next, using an electroless copper-plating process, a copper coating is applied to the hole sides and board surfaces. The steps in the process are:

1. Hydrochloric acid (25%) dip for 1 ½ minutes.
2. Sensitizing in a stannous chloride solution for 1 ½ minutes.
3. Rinse in deionized water.
4. Activate in a palladium chloride solution for 1 ½ minutes.
5. Deionized water rinse.
6. Electroless copper bath: a solution containing copper sulfate, sodium hydroxide, formaldehyde, a chemical reducing agent to hold the copper in solution, and other special additives. The time in solution determines the thickness. The deposition process is:

$$CuSO_4 + 2\,HCHO + 4NaOH \xrightarrow{Pd} Cu + 2\,HCO_2\,Na + H_2 + 2\,H_20 + Na_2\,SO_4$$

7. Deionized water rinse.

The copper layer is built up to about 40 microns with a standard electrocopper deposition process (Section 2.2.6).

The photolithographic process occurs with the resist removed from the area where the tracks are to be placed. This is the reverse of the process described in Section 2.2.3. The exposed track areas are electroplated with either tin or tin/lead to a thickness of about 10 microns (alternatively, gold can be used). This metal is used as a resist in the etching process, which is the next step. Following the etch, the solder is hot air leveled and then the solder mask is screen printed on.

In the case of a multilayer board, composed of alternate layers of copper track and laminate, each of the inner track layers are formed as if they were a single-sided board. No drilling is done. Extreme care must be taken to ensure that they are photoengraved in line with the punched alignment holes. The tracks are attached to a very thin laminate. All the layers are sandwiched together with unetched copper top and bottom layers. Alignment is critical as there can be as many as 40 layers of track. The stack is laminated to form a single multilayer board (Section 2.2.1). The board is now treated as if it were a double-sided, plated through-hole board and processed as shown in Figure 2.13.

Flexible boards are made as single-sided boards. They are normally punched and not drilled. The laminate may be epoxy, fluorocarbon or polyester. These and other examples are given in Table 2.6.

Table 2.6 Selected flexible laminates and their properties

Resin material	Maximum temperature of operation °C	Dielectric constant
Epoxy	125–250	5.0
Fluorocarbon	200–250	2.5
Polyester	150	3.0
Polyimide	400	3.5
Polycarbonate	120–150	3.0
Polyethylene	110	2.2
Polysulfone	160	2.8
Polyproplylene	125–160	2.0

2.4 New materials and technologies

Problems created by the introduction of surface mount techniques to printed circuit boards have given rise to new materials and ideas. An initial problem is the difference in the thermal coefficient of expansion between a ceramic package (5–7 ppm/°C) and that of a laminated board (15–18 ppm/°C). Packages without leads undergoing thermal cycling tend to lift off boards. Although the use of leaded surface mount packages has partly relieved this problem, it is still present. If silicon chips are to be direct bonded to laminates their thermal coefficient of expansion must also be low, typically 5–6 ppm/°C. As a result, new board materials, with better coefficients of thermal expansion, are becoming available.

Another difficulty with conventional laminate is the board dielectric constant. It is not constant but shows dielectric dispersion and is often called the 'Hook' capacity (Doeling & Mark 1978). Boards with lower and more consistent dielectric constants are also appearing. Table 2.7 gives details on some of these materials (*Electronics* 1986; Dance 1986; *Electronic Design* 1988).

An alternative approach to the thermal coefficient of expansion problem is to use flexible substrates (El Rafale 1982*a* and 1982*b*; Winard 1983). The substrates may be totally flexible or only flexible at the surface. In the latter case, an elastomer 50–100 microns thick is coated to the board surface to cushion the thermal mismatch between a leadless ceramic chip carrier and the board. The elastomer is applied by either dipping, screen printing or laminating by heat. It may be a polyimide, epoxy or vitrille rubber material. The process was designed by ITT. A more radical approach is the fully flexible substrate of thin sheets of polyimide with deposited copper layers, developed by Welwyn Electric. Two layer systems can be used which retain the advantages of flexibility and, when provided with clearance holes and a suitable pillared copper or aluminum plate, can be built into a compact unit with good heat sinking (Smith 1983).

Yet another approach is to use a sandwich of metal and plastic to form a composite board (Lassen 1981; Dance and Wallace 1982; Dance 1983; Gray & Elkins 1988; *Electronic Engineering* 1989; *Electronics Weekly* 1989). Texas Instruments and General Electric both report good test results with a copper-clad invar board. A useful advantage of metal core printed circuit boards is that they not only match the thermal temperature

Table 2.7 Properties of new laminate materials compared to an epoxy/glass laminate

Material	Dielectric constant	Dissipation factor	Thermal coefficient expansion ppm/°C	
			X and Y axis	Z axis
PTFE	2.1	0.0004	224	224
PTFE/micro glass	2.2	0.0008	24	261
PTFE/ceramic	2.8	0.002	19	24
Polyimide/quatz	3.35	0.005	6–8	34
Polyimide/Kevlar	3.6	0.008	3.4–6.7	83
Polyimide/glass	4.5	0.010	11.7–14.2	60
Epoxy/glass	4.8	0.022	12.8–16	189

coefficient of expansion and provide mechanical rigidity, but they also allow good heat sinking. Many high speed VLSI circuits need to dissipate large amounts of heat and the removal of this heat is a growing problem. Boards with heat pipes, solid metal cores, and hollow boards with liquid cooling are alternatives that are becoming available.

A problem with card technology is the large difference in track pitch a card has compared to that of an integrated circuit. If the card track spacing and width can be reduced to 120 microns or less, then not only could chips be gang bonded directly to the board, but increased board track density would significantly reduce the size. The use of lasers to drill via holes 175 microns in diameter on 500 micron pads (*Circuits Manufacturing* 1986) and electron beam lithography for printed circuit boards, allowing track width and spacing down to 60 microns (*Electronics* 1987), are two approaches adopted to overcome this problem.

Flexible/rigid boards are also possible so that cards can fold like books into a more compact space (Rigling 1988). The flexible section is typically 50 microns thick for strength, using either polyester or polyimide film, acrylic, epoxy, or polyester adhesives 12–25 microns thick cementing on the ½ or 1 ounce copper track.

Yet another innovation is the molding of printed cards using injected molded thermoplastics. In effect, the case or box housing the electronics becomes the printed wiring board (Ganjec 1986; Lyman 1988; Whelan 1990). This has only become possible with the introduction of surface mount technology. Several additive processes are available for placing the track on the plastic. These include polymer thick film printing, conductor channels and plated-up copper.

2.5 Questions

1. What are the essential components of a bare laminated fiberglass board? Describe how such a board is made.
2. Why is cleanliness so important in the manufacture of boards? What cleaning steps would you implement to prepare a copper laminate for:

   (a) photo engraving
   (b) plating?

3. Draw a flow diagram of the steps involved in making a multilayer board.
4. Compare wet and dry film photoresist technologies by listing the advantages and disadvantages of each type.
5. Why are hot air leveling systems used in the manufacture of tin or solder plated copper printed circuit boards? Explain the operation of a hot air leveling system.
6. What is electroplating? How is it employed in the manufacture of electronic cards?
7. Prepare a list of possible applications for flexible printed circuit cards.
8. List ways in which heat may be removed from chips mounted on a printed circuit board.

2.6 References

Bosshart, W. C. (1983), *Printed Circuit Boards*, Tata McGraw-Hill, New Delhi.

Circuits Manufacturing (1986), 'Laser vias beat mechanical vias', *Circuits Manufacturing,* **26** (August), p. 15.

Coombs, C. F. (ed) (1967), *Printed Circuits Handbook*, McGraw-Hill, New York.

Dance, F. J. (1983), 'Low thermal expansion rate clad metals for chip carrier applications', *Electronics Show and Convention*, May 10–12, pp. 1–5.

Dance, F. J. (1986), 'Printed wiring boards: A five year plan', *Circuits Manufacturing*, **26** (June), pp. 23–36.

Dance, F. J. & Wallace, J. L. (1982), 'Clad metal circuit board substrates for direct mounting of ceramic chip carriers', *Circuit World*, **9** (January), pp. 47–9.

Doeling, W. & Mark, W. (1978), 'Getting rid of Hook: The hidden PC-board capacitance', *Electronics*, **51** (12 October), pp. 111–17.

Electronic Design (1988), 'PC boards hot-wired for faster more flexible service', *Electronic Design*, **36** (18 February), pp. 29–36.

Electronic Engineering (1989), 'Contraves make PCB packing density progress', *Electronic Engineering*, **61** (753), p. 23.

Electronics (1986), 'A speedy board laminate suits surface mounting', *Electronics*, **59** (27 November), pp. 79–80.

Electronics (1987), 'E-Beam could slash turn around time for ultrafine-line PC boards', *Electronics*, **60** (3 September), p. 50.

Electronics Weekly (1989), 'Substrates are coming in from the cold', *Electronics Weekly*, (1476) September 13, p. 20.

El Rafale, M. (1982*a*), 'Chip-package substrate cushion dense, high speed circuitries', *Electronics*, **55** (14 July), pp. 134–41.

El Rafale, M. (1982*b*), 'Chipstrate interconnection substrates for advanced electronic systems', *Electrical Communication*, **5–7** (2), pp. 142–9.

Ganjec, J. (1986), 'Mould your own', *Circuits Manufacturing*, **26** (June), pp. 39–50.

Gray, F. & Elkins, M. (1988), 'Polymer-on-metal multilayer boards', *Circuit World*, **14** (3), pp. 12–21.

Harper, C. A. (ed.) (1970), *Handbook of Materials and Processors for Electronics*, McGraw-Hill, New York.

Karpel, S. (1988), 'Printed circuit boards for industry', *Tin and Its Uses*, International Tin Research Institute, (155) pp. 5–8.

Lea, C. (1988), *A Scientific Guide to Surface Mount Technology*, Electrochemical Publications, Ayr, Scotland.

Lassen, C. L. (1981), 'Use of metal core substrates for leadless chip carrier connection', *Electronic Packaging and Production*, **21** (March), pp. 98–104.

Lyman, J. (1988), 'Grooved boards challenge conventional PC boards', *Electronics*, **61** (26 May) pp. 77–8.

Rigling, W. S. (1988), *Rigid-Flex Printed Wiring Design for Production Readiness*, Marcel Dekker, New York.

Smith, K. (1983), 'Combo interconnect reduces stress on chip carriers', *Electronics*, **56** (10 March), pp. 85–6.

Smith-Vargo, L. (1987), 'Process windows widen on aqueous resists', *Electronic Packaging and Production*, **25** (May), pp. 65–7.

Spitz, S. L. (1987), 'From schematic to PCB with phototools', *Electronic Packaging and Production*, **27** (5), pp. 56–8.

Whelan, M. (1990), 'Three-dimensional moulded printed circuits', *Circuit World*, **16** (2), pp. 25–9.

Winard, H. (1983), 'PC-type substrate handles dissipation', *Electronic Design*, **31** (31 March), p. 175E.

3 Solder and the soldering process

3.1 Introduction

As a first order approximation, the reliability of a product is inversely proportional to the number of component interconnection joints. Thus the weakest link in a product is the connection and care must be taken to employ methods that achieve the highest joint reliability possible.

A satisfactory joining method should:

1. be simple and repeatable
2. provide adequate mechanical strength
3. give joints of low electrical resistance
4. offer a reversible process, so that joints can be undone
5. not damage the components being joined.

There are four common metallurgical methods employed for joining conductors. They are: (1) soft soldering, (2) hard soldering, (3) brazing, and (4) welding. In the electronics industry, only soft soldering and welding are used because the other two processes normally require excessively high temperatures to form joints.

In this chapter the soft solder method will be discussed because this process has become, and remains, the dominant joining process in the electronics industry. Welding and the other classes of electronic joining techniques will be discussed in Chapter 4.

3.2 Electronic grade solder

Soft solder is the alloy of two elements, tin and lead, which melt at a low temperature to make a joint. A flux is normally added to the joining process, to provide final cleaning of the component leads, thereby enabling good wetting by the solder.

3.2.1 Solder alloy

Figure 3.1 (RAAF 1972) shows the metallurgical phase diagram for tin/lead mixtures. The horizontal axis specifies the ratio of the two materials by weight, while the vertical axis shows temperature. Thus, on the left hand side, there is pure lead. Moving to the right side, the percentage of lead decreases and tin increases, so that on the far right hand side the material becomes pure tin.

Table 3.1 Standard tin/lead alloys and their applications

Alloy tin/lead	Resistance at 25°C micro ohms per meter	Melting temperature °C	Recommended soldering temperature °C	Uses
60/40	0.150	188	248	High quality electronics
50/50 45/55 40/60	0.158 — 0.171	212 224 234	272 284 294	Electrical equipment, batteries
30/70 20/80	— —	255 275	315 335	Fuses, motors, dynamos, lamps

Figure 3.1 shows that the melting points for pure lead and tin are 327.5°C and 232°C respectively. Following down the liquid line, the alloy combination that gives the lowest melting point of 183°C is a tin/lead ratio of approximately 63% tin and 37% lead. This is also the eutectic point, where the alloy changes from solid to liquid without going through a plastic phase (often referred to as pasty). For components that are temperature critical, alloy combinations near the eutectic mix appear most suitable, giving the lowest melting temperatures and a rapid transition direct from liquid to solid.

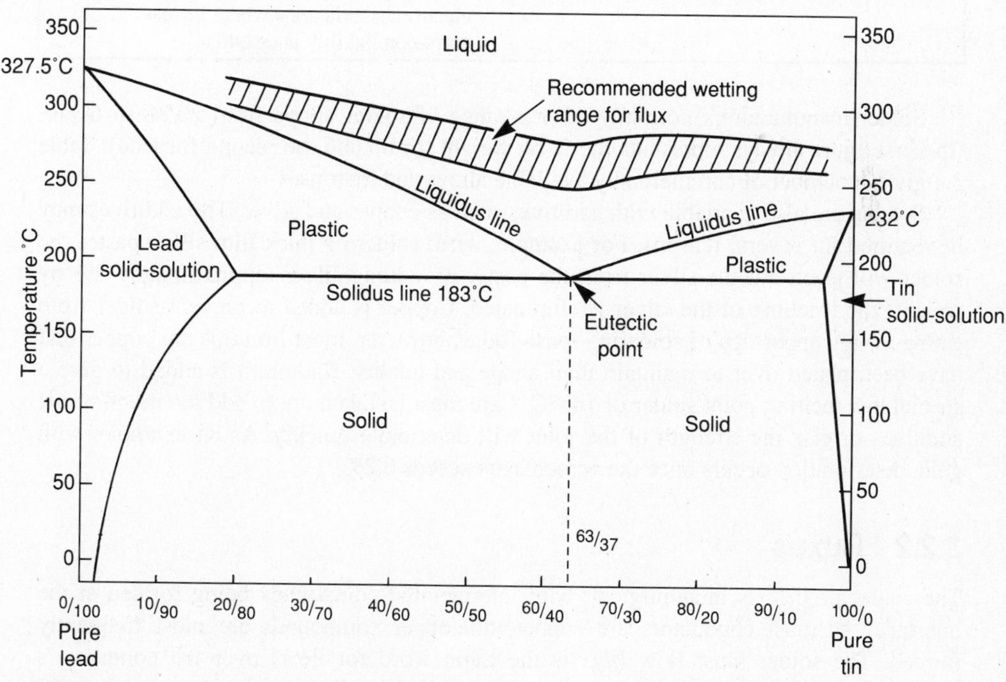

Figure 3.1 Phase diagram for tin/lead solder

Table 3.2 Examples of flux types

Chemical basis	Applications
Zinc chloride with or without other inorganic halides Inorganic acids such as hydrochloric acid	Soldering and 'tinning' of ferrous and non-ferrous metals, in general engineering, sheet metal work and plumbing. Flux residues are normally removed.
Phosphoric acid and derivatives	Soldering of stainless steel. In some circumstances the residues from these fluxes need not be removed.
Halides of organic compounds (including hydrazine)	Soldering and 'tinning' of non-ferrous metals in general engineering and sheet metal work. Flux residues are less corrosive and less electrically conductive than zinc chloride or hydrochloric acid and often may not need to be removed. On electronic assemblies, their removal is recommended.
Organic acids	Soldering of lead, pewter and copper base alloys.
Activated resins: (a) Activators containing a halide (b) Activators not containing a halide	Soldering and 'tinning' of non-ferrous components in the electrical and electronics industries. Suitable for use with high-speed mechanized soldering systems.
Non-activated resins	Soldering of copper in the electrical and electronics industries where lowest corrosion liability is essential.

Solder manufacturers normally offer a range of solder alloys from 20/80 to 60/40, (the first figure always represents the % by weight for tin and the second for lead). Table 3.1 gives a number of commercially available alloys and their uses.

Solders are also available with additives such as copper and silver. The additives may be required for several reasons. For example, when soldering thick film silver pastes the solder will leach out the silver from the paste. By adding silver (approximately 2% by weight) the leaching of the silver is eliminated. Copper is added to prevent solder from eating away copper tips of soldering ions. Today, however, most iron tips are copper that have been plated over to maintain their shape and quality. Cadmium is added to give a special low melting point solder of 145°C. Care must be taken not to add too much of the additives or else the strength of the joint will deteriorate quickly. As an example, with gold, deterioration occurs once the percentage exceeds 0.2%.

3.2.2 Fluxes

The solder process is metallurgical, with intermetallic compounds being formed at the interface. As most conductors are copper, tin/copper compounds are most frequently formed. The solder must flow (*flux* is the Latin word for flow) over the conductor's surface and wet it to achieve a good joint. If the balance between surface tension and wetting is incorrect, the solder will produce incorrect joint wetting angles and joints with poor mechanical and electrical properties. In the worst case of almost no wetting, solder globules form.

To achieve good wetting it is essential that the joint materials are clean. Oxides must be removed mechanically and chemically before starting the soldering process. To maintain this cleanliness even at the elevated temperatures necessary for the joining process, fluxes are used. Their function is to remove the remaining thin oxide layers and cover the surface to prevent further oxidization. Therefore, at the elevated process temperature, all fluxes are active and corrosive.

There is a wide range of fluxes available, as shown in Table 3.2, and the one selected often depends upon the materials to be soldered. If nichrome resistor wire is to be soldered, a strong acidic flux such as hydrochloric acid must be used. Zinc chloride salts, though common in the plumbing industry, are frowned upon in the electronics industry because they are highly corrosive, even at ambient temperatures. What is required is an inert flux, so that any residue will not cause corrosion at a later date.

There are two basic approaches in the electronics industry. Some manufacturers prefer to employ strong acid fluxes to ensure printed circuit boards are clean and good joints result. Since any residue can cause long-term problems of corrosion, stringent washing procedures have to be established and tests made to ensure that the amount of residual flux material is well below the limits set. By far the most common method is to use activated resin fluxes. Here the flux is inactive at ambient temperature and only becomes active at the elevated soldering temperature. They are usually resin based and

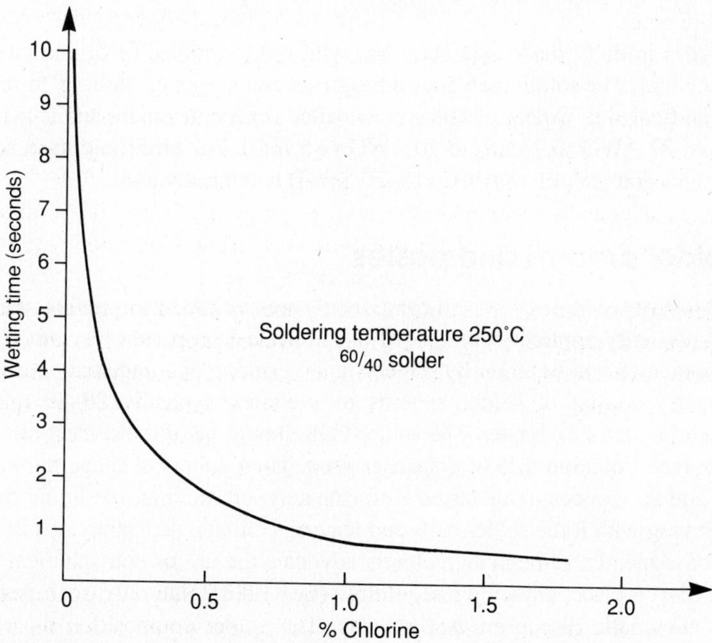

Figure 3.2 Effect of a chlorine activator on removing the oxides of copper and allowing wetting

may use a halide activator to improve their tarnish-removal properties. The resin (or rosin) is usually colophony, an extract from the sap of some pine tree species, dissolved in alcohol. Synthetic equivalents are also available. Halide activators may be dimethyl-ammonium chloride or diethyl-ammonium chloride. The time needed for solder to wet a surface depends upon the number of chlorine ions present. Figure 3.2 shows the influence of the activators in cleaning a copper board that has been oxidized at 150°C for 1 hour.

The resin is categorized by the amount of activator. Thus R is pure resin, RMA is resin mildly activated and RA types are resins highly active due to strong ionic activators to clean away oxides.

Because resin fluxes are inactive at room temperatures, they are sometimes left on a board. When they are to be removed, organic solvents are needed. By adding a saponifier to the flux they can be semi-aqueous (water and detergent) or aqueous cleaned (Lea 1988).

The temperature to which the flux is heated prior to soldering is critical. At too low or too high a temperature the flux will not clean the copper adequately. The ideal temperature range is about 250 to 275°C. Figure 3.1 includes the recommended temperature range for resin-based fluxes to achieve good wetting. Thus, when soldering, the temperature of the iron or molten solder must be higher than the solder melting point.

3.2.3 Solder for hand soldering

To assist in the process of hand soldering (either to assemble or rework) the flux is often combined with the solder by placing it in channels within the solder itself. Figure 3.3 shows two core types.

The solder is initially made as a short bar with either molded or drilled channels to incorporate the flux. The solder then goes through several stages of 'pulling' to reduce the diameter to the final size. Solder of this type is called resin core solder and is available in diameters from 22 AWG (0.7 mm) to 10 AWG (2.5 mm). For electronic hand soldering, 0.45 inch (1 mm) diameter or 18 AWG (18–20 SWG) is normally used.

3.2.4 Solder creams and pastes

With surface mount technology, where components are soldered using reflow techniques, the solder is normally applied to the board or substrate in a paste or cream form. The method of application can be either by screen/stencil printing or a pneumatic dispenser.

Solder pastes consist of solder spheres or globules, typically 20–80 microns in diameter, mixed with a flux binder. The solder balls should be of the correct diameter for the mesh size (see Equation 6.1) or dispenser used. Their spherical shape minimizes the surface area and so reduces oxidization. Unfortunately, on melting, the liquid flux flows outwards, carrying with it the solder balls and leaving behind a deficiency of solder called a 'slump'. Consequently, some manufacturers advocate the use of non-spherical particles to lock the solder in place, allowing less outflow (Lea 1988). Naturally, such pastes more readily clog pneumatic dispensers and screens. The solder composition must be well controlled because most impurities deteriorate joint quality. Copper, iron and gold form intermetallic compounds, causing the solder to become gritty and brittle. Aluminum, cadmium, and zinc promote oxidization of the solder surface, causing the solder to be

Table 3.3 Selection of solder paste types available

Composition tin/lead	Melting point °C	Viscosity CPS (23°C Brookfield Viscometer)	Chlorine content	Uses
63/37	183	30.10^4	0.2	Electronic printed circuit boards
62/36 2 Ag	179	30.10^4	0.2	Hybrid integrated circuits, silver electrodes
5/92.5 2.5 Ag	280	20.10^4	0.2	High temperature soldering and hybrid integrated circuits
42/42 14 Bi 2 Ag	160	30.10^4	0.2	Low temperature soldering and hybrid integrated circuits

sluggish. Sulfur causes tin and lead sulfides to form, which inhibit soldering (ISHM 1984). The binder not only includes the activated flux, but also organic solvents, thickeners, and lubricants to determine paste rheology. With screen or stencil printing of solder, rheology is critical to ensure excellent print definition. Smaller solder balls allow improved print resolution, but increase the solder surface area. This results in considerably more oxidization of the solder which means the flux must work harder to prevent solder balling. Various grades of paste are available, some more suitable for screen printing and others for application via a pneumatic syringe dispenser. Table 3.3 provides further information on such pastes.

3.3 The soldering process

3.3.1 Preparation

As previously stressed, quality solder joints can only be achieved under clean conditions. Boards, component leads, and soldering equipment must be carefully cleaned both before and after the soldering operation. In the case of bare oxidized copper, mechanical

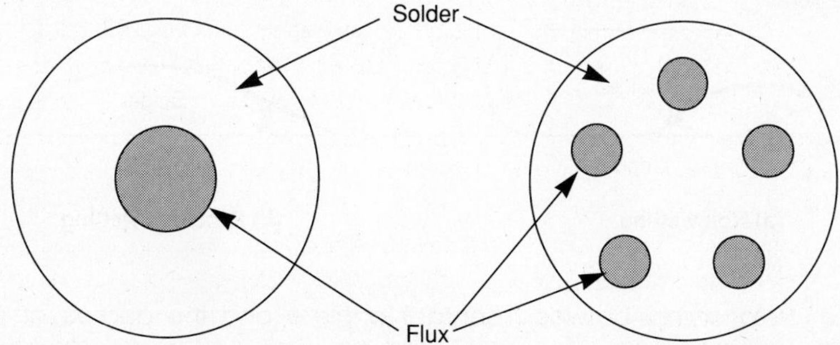

Figure 3.3 Examples of configurations of resin flux incorporated within the solder

scrubbing with a fine abrasive may be required. For small board quantities and hand soldering this can be done by hand using a coarse eraser. A circular motion should be used on the parts to be soldered so that scratches made in the copper will help contain the solder and not allow it to flow along the copper and away from the joint. Following mechanical scrubbing, a chemical clean should be carried out. Special organic solvents are available for this process. Chlorinated fluorocarbon (CFC) should not be used because of its long-term impact on the ozone layer. Alternatives such as hydrochloridefluorocarbon (HCFC) and hydrofluorocarbon (HFC) are now becoming available. These are believed to have less effect on the environment.

3.3.2 Soldering variables

Three main variables must be controlled to achieve repeatable quality joints:

1. Temperature: there is both a minimum and a maximum temperature, with 250–270°C being the preferred range for 60/40 solder.
2. Time that heat is applied: again, there is a minimum and maximum time limit, usually ranging between 5–7 seconds.
3. Flux application.

Depending on whether hand or automatic soldering methods are used, the flux and the solder may be applied separately or together. The temperature must be correct for the flux to be activated to remove any remaining oxide and apply a protective coating. The heat source not only melts the solder, but boils away the resin and flux solvents so that the solder flows over, wetting the copper and giving a low angle of wetting ($\theta \rightarrow 0°$) as shown in Figure 3.4. For $\theta < 40°$ wetting is considered adequate (Manko 1964).

The temperature applied should not be excessive or the joint mechanical strength deteriorates. The wetting temperature range of 250–270°C is just above the temperature of maximum strength and therefore an acceptable temperature range. Above 300°C there is a rapid deterioration in mechanical strength. All components being soldered have a maximum temperature/time profile, and if an excessively high soldering temperature is

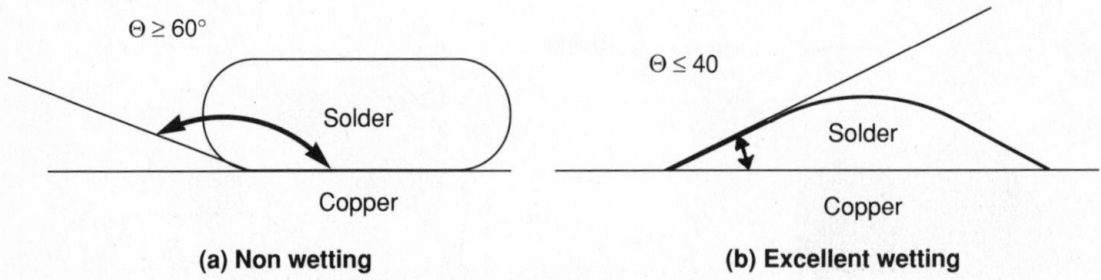

Figure 3.4 Relationship between contact angle θ and the degree of surface wetting. For $\theta < 40°$, wetting is considered adequate while for $\theta \geq 60°$ the joint is unacceptable

applied or a lesser temperature applied for too long (a few seconds in some instances) the component is damaged. Damage can be caused to the:

1. printed circuit board—tracks lifted, board blistered
2. flux—degradation in performance
3. electronic component—cracked package seals, increased leakage currents, thermal shock to ceramic components.

The solder on the completed joint should have a smooth, shiny surface. Should it have a gray matt finish, too much heat was applied. If there are indentations or pin holes, there may be trapped gases or contaminants. If a line or ridge, there may have been a different cooling profile for each part of the joint. These defects are illustrated in Figure 3.5. In all cases the joint strength and reliability will be poor. If the board is gold plated, sufficient gold may have leached out to poison the solder and again reduce its quality. Finally, the strength of a joint depends on the quality of wetting and not the amount of solder used. Excess solder simply adds weight and increases costs.

3.3.3 Soldering processes

The actual soldering process may be manual or automated. It may involve the application of solder or reflow of solder. Figure 3.6 illustrates the options. Reflow methods are normally used for surface mount components and the solder additive processes for leaded components (component leads are inserted through the board and soldered to the underside). These joints may be clinched or rigid.

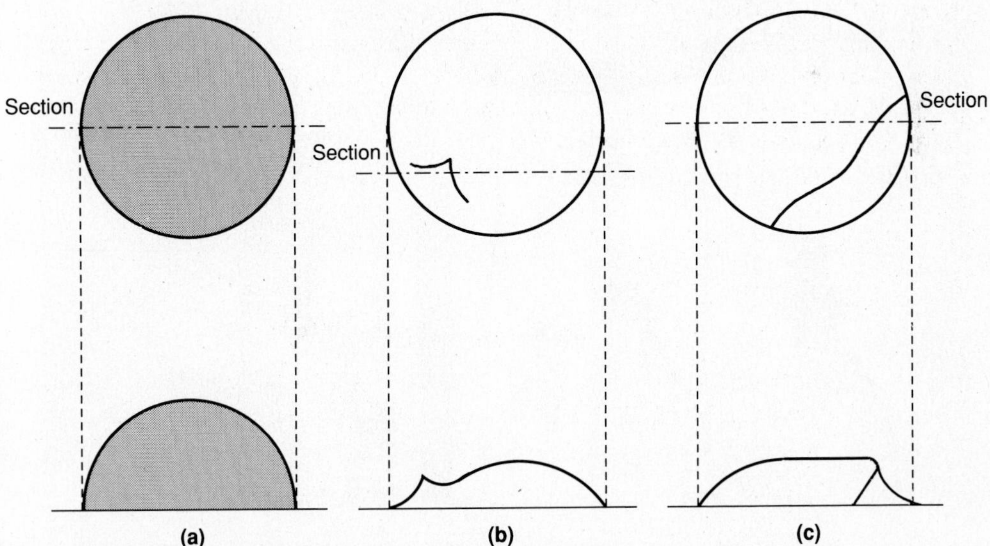

Figure 3.5 Solder defects: (a) overheat and too much solder, (b) trapped impurities, and (c) each half solidified at a different time

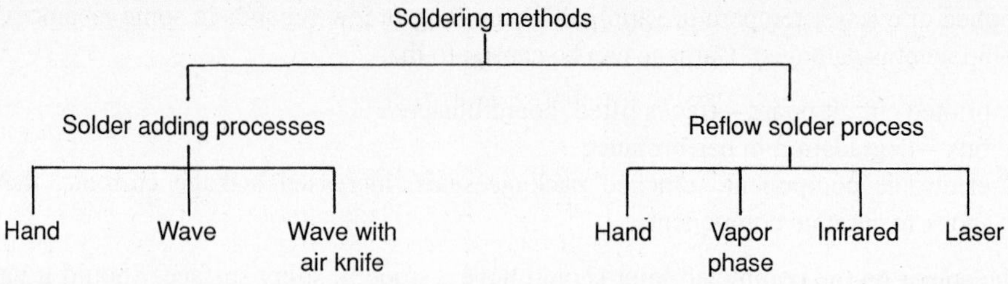

Figure 3.6 Soldering methods used in industry

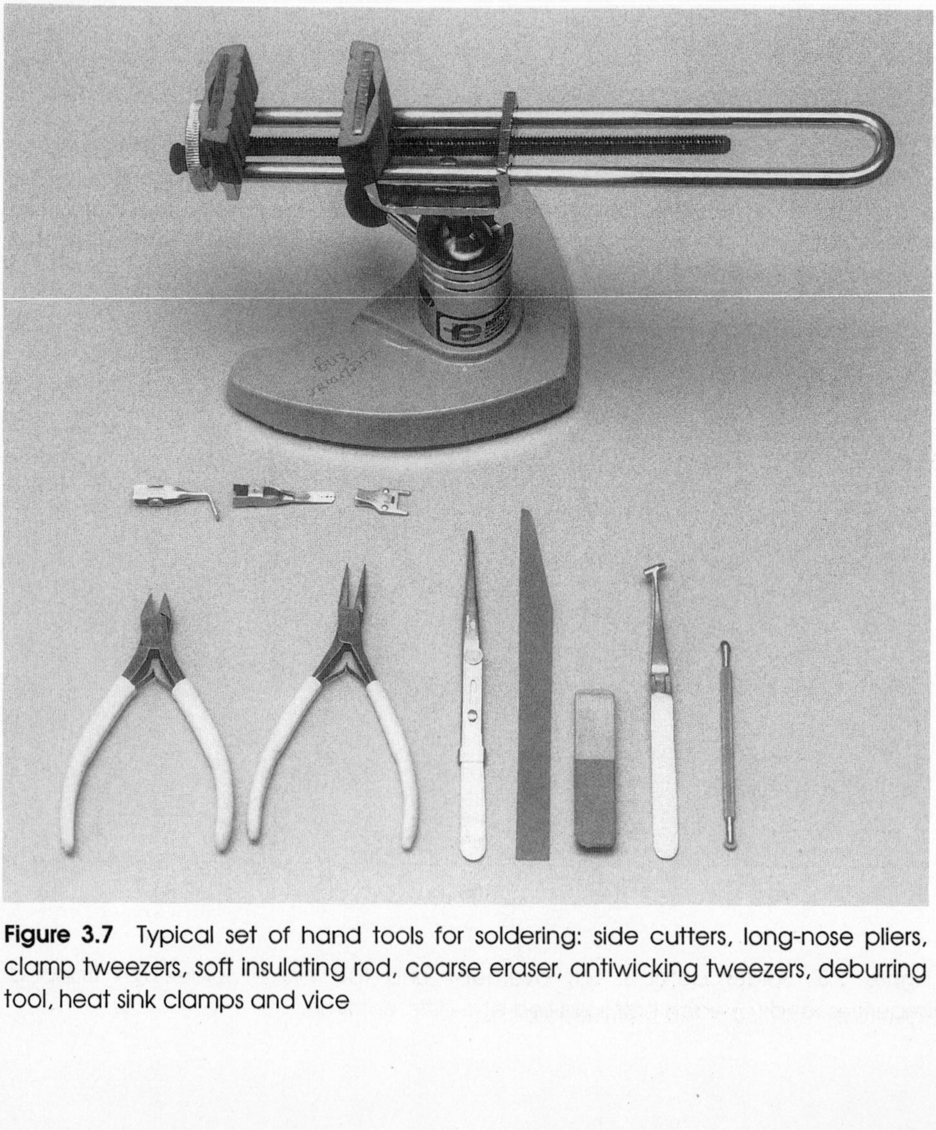

Figure 3.7 Typical set of hand tools for soldering: side cutters, long-nose pliers, clamp tweezers, soft insulating rod, coarse eraser, antiwicking tweezers, deburring tool, heat sink clamps and vice

During the changeover period from through-hole to surface mount methods, wave soldering techniques have been adapted to solder surface mount components. This delays the capital outlay to purchase new soldering equipment.

Before each of the methods shown in Figure 3.6 is discussed, including a description of the equipment used and the soldering process steps, it should be pointed out that adequate safety must be ensured. Exploding hot solder, volatile toxic and build-up of corrosive vapor are to be avoided and, should an accident occur, operators must have adequate protection.

3.4 Hand soldering

3.4.1 Hand tools

To assist with hand soldering, several hand tools, other than a soldering iron, are essential. They are:

- side cutters
- long-nose pliers
- clamp tweezers
- semi-rigid insulating chisel end rod
- coarse eraser
- antiwicking tweezers
- heat sink clamps
- vice.

They are shown in Figure 3.7. In keeping with the size of modern electronic components, the tools should be small, but not so small that they feel uncomfortable in the hand.

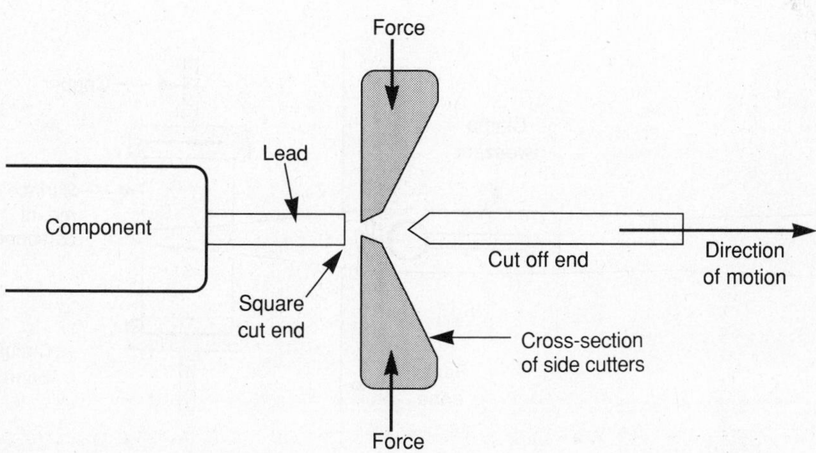

Figure 3.8 Cutting a lead with side cutters

Side cutters should be flush on one side so that the wire is cut with a square end on the component side. The tapered edge of the cutters imparts a horizontal force to the cut end and care should be taken to ensure that the cut end does not fly into equipment or anyone's eyes (see Figure 3.8).

The long-nose pliers should be plain and have no serrations that can damage the lead and expose the copper underneath the tin plating. The corners of the pliers should not be sharp and, if they are, must be given a slight curve using emery paper. The pliers are used for shaping leads and forming clinched joints. Any rough edges will cut into the lead and weaken it as well as exposing the lead material which may oxidize. This is illustrated in Figure 3.9.

The clamp tweezers are useful for holding surface mount components in place while soldering. This is illustrated in Figure 3.10.

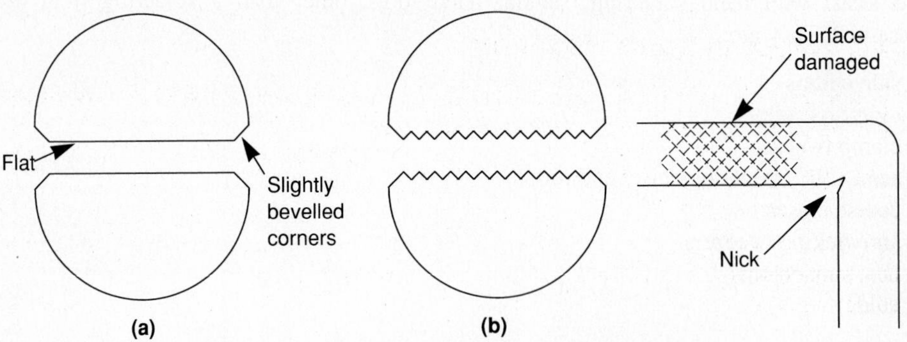

Figure 3.9 Correct (a) and incorrect (b) cross section of long-nose pliers. Possible defects made on leads with incorrect pliers

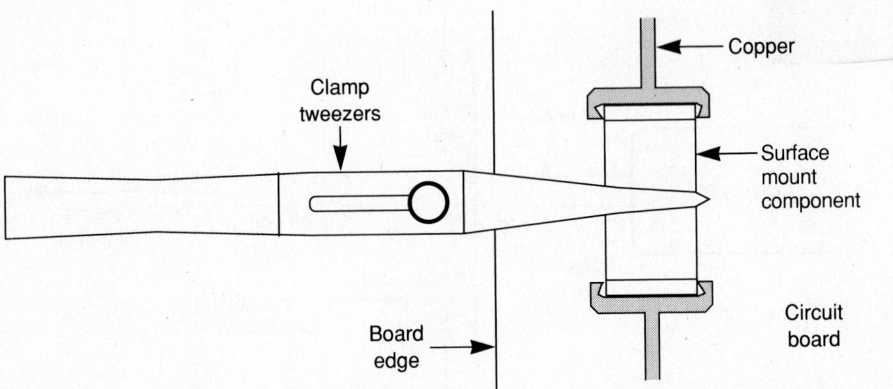

Figure 3.10 Using clamping tweezers to hold a component in place for hand soldering

Finally, a plastic rod is used for flattening component leads onto the board, particularly clinched leads (see Figure 3.11). The material must be soft enough so that it does not damage the component or board. An offcut of thick epoxy printed circuit board with the copper removed is satisfactory.

Many other hand tools are available and, depending upon throughput, may be useful to have. These include:

- Wire stripping tools to remove any insulation and uncover bare and undamaged wire ready for soldering.
- Component lead forming tools, varying from simple jigs that can be preset to consistently bend leads of axial lead components through to hand machines that grip components and perform lead bending. Figure 3.12 illustrates these lead forming tools.
- Deburring tool (Figure 3.7) to remove, before soldering, any rough copper edges caused by the drilling of holes in a board.

3.4.2 Steps in hand soldering

It is not difficult to achieve consistency in solder joint quality using hand methods, providing several basic steps are adhered to. These are:

- soldering iron design
- operating temperature
- cleaning processes
- components to match the board layout
- soldering steps.

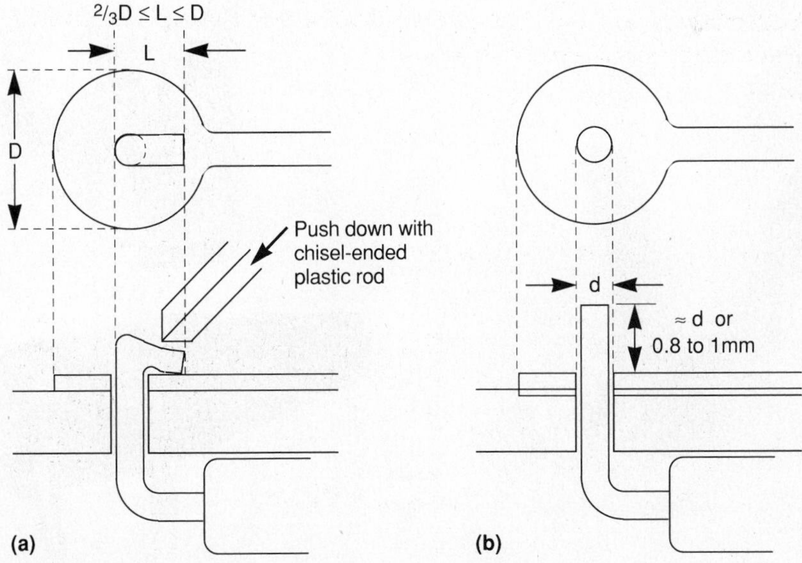

Figure 3.11 Prepared leads for soldering: (a) clinched and (b) rigid joints

The soldering iron must be temperature controlled, with the temperature sensing element as close as possible to the tip end. Depending upon the size of the component being joined, the iron must have sufficient thermal capacity so that the set and working temperatures are approximately the same. The iron should be isolated so that there is no large leakage voltage appearing on the tip to cause damage to components. Two types of soldering iron are commonly used. The first is a fully closed loop, electronically controlled unit which has a temperature sensor as near as practical to the tip so that the true tip temperature is monitored and controlled. Some models also provide a readout of tip temperature. The temperature is normally maintained at ± 2°C. The second and cheaper type employs special set temperature tips that magnetically switch power at the control material Curie point. Temperatures are typically held within ± 6°C. The available tip temperatures are 315°, which is normal for the assembly of electronic cards, 370° and 430°C. Figure 3.13 shows examples of soldering irons.

Soldering iron tips are normally copper to provide good heat transfer, but are plated to prevent the solder dissolving them. Iron, on the other hand, is not 'attacked' by solder, so iron plated copper tips are normally used today. Unfortunately, iron is not readily wetted by solder, so the tip is further covered by a nickel and/or chrome plating to give a hard outer surface that is wetted by the solder. With time, the nickel will dissolve away and the bit must be replaced. Excess solder and burnt flux can be removed from the tip by wiping it on a water-soaked sponge. If the burnt flux is too thick the iron temperature must be set as high as possible before plunging it into the wet sponge. Tips come in various shapes and sizes, including conical and chisel. With a chisel tip the width should be approximately two-thirds the copper pad diameter. Figure 3.14 shows a section of a tip and alternative shapes.

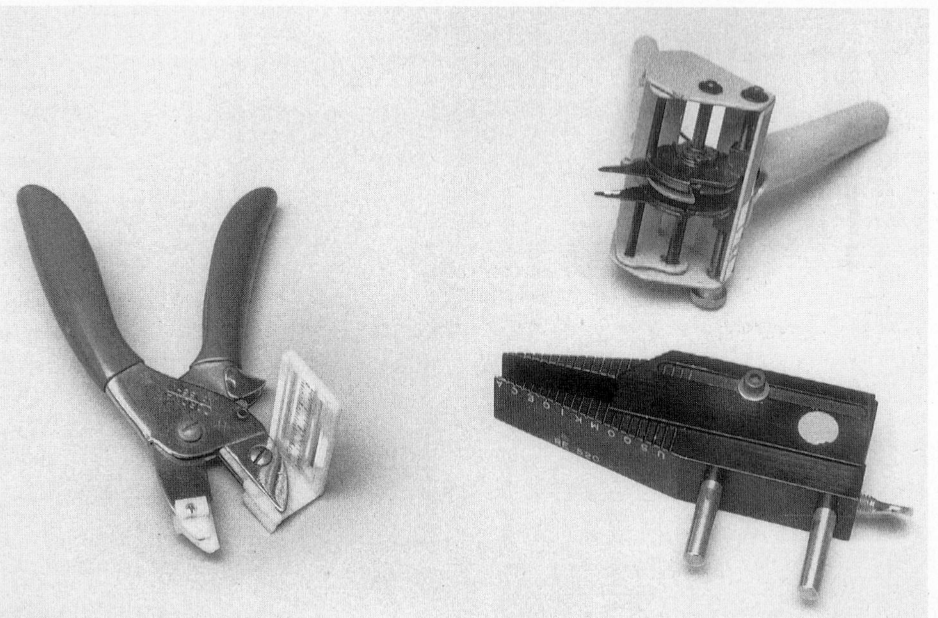

Figure 3.12 Examples of lead forming tools and jigs

For normal iron operation with 60/40 solder, the soldering iron temperature should be set to about 300–320°C which allows joints to be made at temperatures of 250–275°C.

Board cleaning has already been discussed in Section 3.3.1. The solder should not be stretched or breaks in the flux core may occur, leaving insufficient flux for the soldering process. The solder should always be cut, providing a clean, flux-filled core. It should also be wiped with a solvent to remove any oxide from the outer surface.

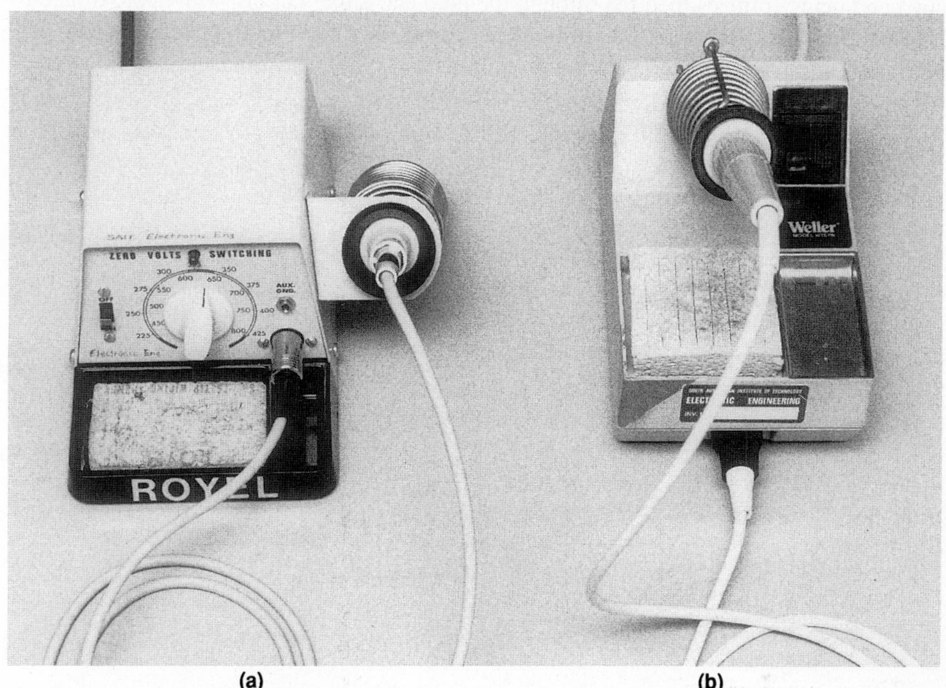

Figure 3.13 Examples of hand soldering irons: (a) a fully controlled unit, and (b) using special magnestat tips

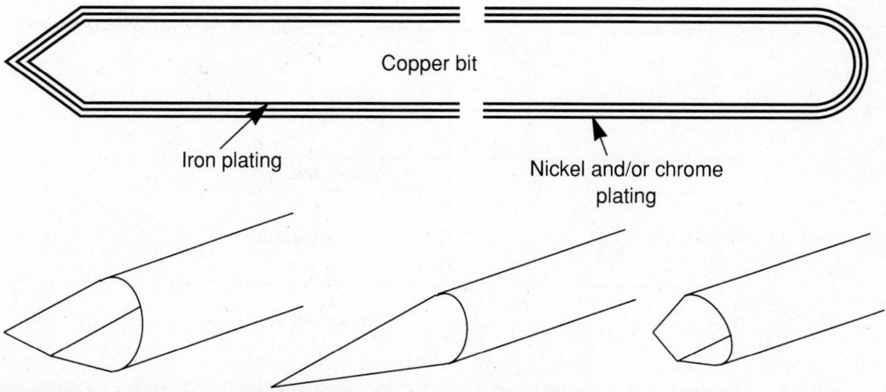

Figure 3.14 Soldering iron tip construction and shapes

Component leads and solder terminations should be cleaned with a solvent to remove any grease or oxide. Finally, the soldering iron tip should be wiped clean on the water sponge and re-tinned. This solder helps form the initial heat bridge between the copper track on the board and the component.

In the case of leaded components, the lead is first formed using the long-nose pliers (to prevent the component being damaged, always grip the lead on the component side of any bend being formed) then fed through the hole and either cut off, with approximately 20–25 mil (0.8–1 mm) projecting through the board for a rigid joint, or for clinched, cut off to a length somewhere between two-thirds to a full pad diameter, as illustrated

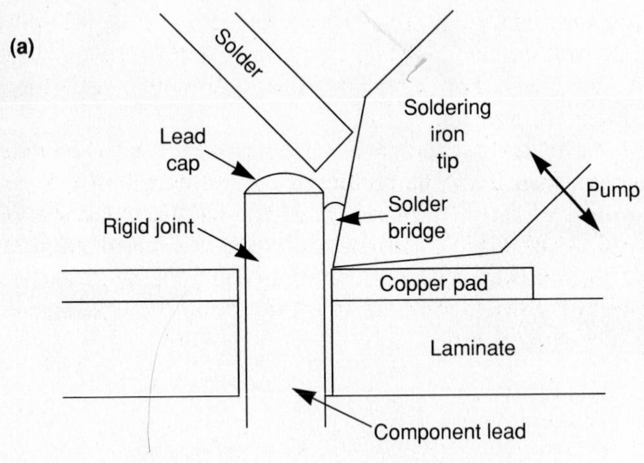

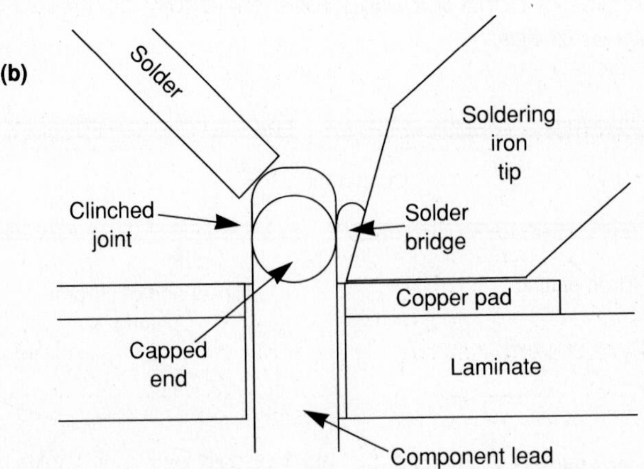

Figure 3.15 Making (a) a rigid and (b) a clinched solder joint with a hand-held soldering iron

in Figure 3.11. The chisel soldering tip, after tinning, is placed against the lead and the copper pad and just sufficient solder fed on to form a heat bridge. Cap the bare end of the cut lead. Run the solder around the outside edge of the pad and, finally, add any additional solder to fill the joint. Care must be taken that excess solder is not added. Pumping the iron up and down a little ensures that solder flows under it, completely wetting all the copper and component lead. This is illustrated in Figure 3.15. Before inspection, the joint must be cleaned with a solvent to remove any residual flux.

For surface mounted components, the pads on the board must first be tinned. Using an artist's small paint brush, flux is brushed over the pad. With a freshly cleaned and tinned iron, new solder is melted onto each pad, just enough to cover the pad evenly. Next, the component is correctly positioned and held in place with either a clamp or a drop of epoxy glue under the component. The tinned chisel iron tip is positioned against the component and the pad, a small heat bridge is formed and the solder allowed to reflow. For components with many leads it often pays to solder corner leads first to ensure the component is positioned correctly.

It has been assumed that the surface mount component has come pre-tinned. Should this not be the case, then it can be pre-tinned by first chemically cleaning the leads, brushing on flux and then dipping into molten solder in a small solder pot. The solder pot is prepared by slightly overfilling, the meniscus holding the solder back in the pot. With a dampened tissue, wipe away any dross (predominantly metal oxides) on the solder surface just prior to the insertion of the lead for tinning. Protective glasses should be worn.

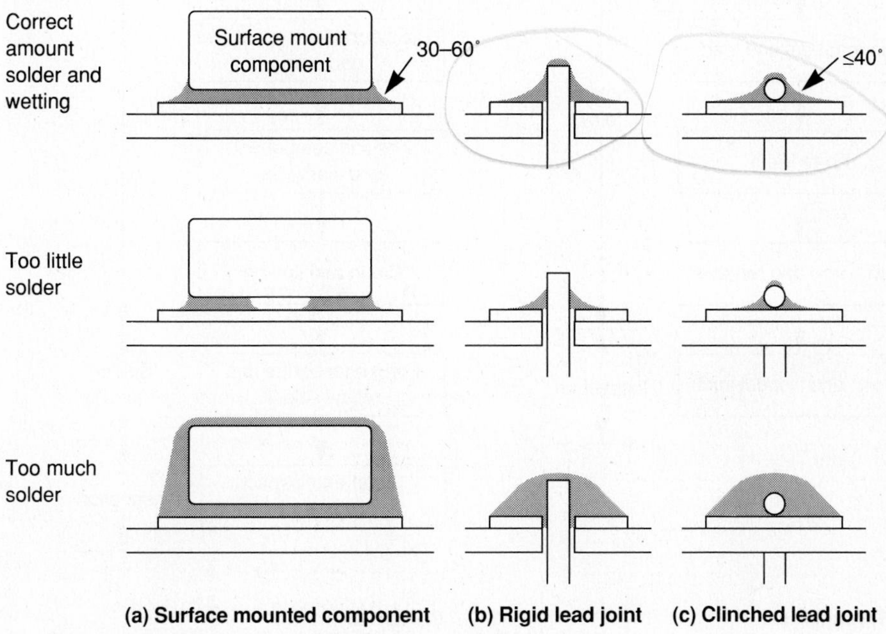

Figure 3.16 Examples of hand soldered joints: (a) surface mounted component, (b) rigid lead joint, and (c) clinched lead joint

Figure 3.16 shows examples of leaded and surface mount soldering. Where too much solder is applied it is difficult to see the component/copper interface to establish if a metallurgical joint has been correctly formed.

In summary, the steps involved in hand soldering of joints are shown in Figure 3.17.

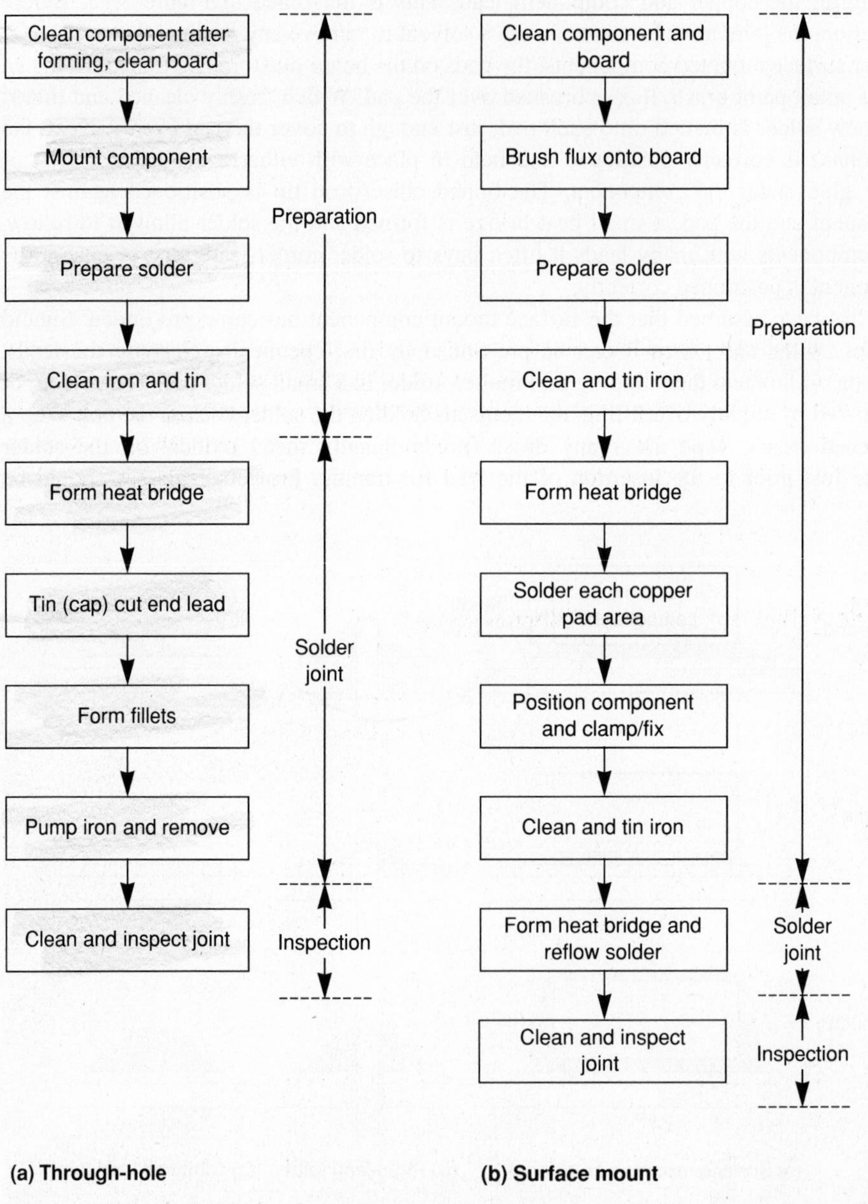

Figure 3.17 Flow diagram for making (a) through-hole, and (b) surface mount solder joints by hand

3.5 Wave soldering

Wave soldering methods were first developed in 1956 (Dummer 1978). Molten solder in a bath is pumped to form a wave that impinges on the bottom of a moving and loaded printed circuit board. Over the years there have been considerable improvements, including inclined conveyor belts, fluxing and preheat stations, oil included in the solder wave and the shaping and number of solder waves. Figure 3.18 shows a typical unit of today. The object is always to improve the quality of the joints and reduce the number of defects. Faults occurring on wave soldering systems can normally be classified as one of three types.

1. Insufficient wetting: often due to shadowing of components preventing flux from reaching tracks and leads. These are also called solder skips.
2. Bridges: where there are shorts between two adjacent tracks.
3. Icicles: so named because of their appearance, namely long, irregular, spindly pieces of solidified solder hanging from the underside of the board.

To reduce these problems, printed circuit boards that are to be soldered are usually screen printed with a solder mask. This is an epoxy material that protects those areas of the board where soldering is not required. Further, component leads are normally guillotined longer than the 20–25 mil (0.8–1 mm) used with hand soldering (Figure 3.11b), typically 1.5–2 mm in length. Figure 3.19 shows the important features of a wave-soldering machine. An inclined conveyor belt (5 or 6° inclination, speeds continuously variable up to 100–150 inches (3 or 4 meters) per second) first takes boards, mounted in a frame or pallet, across a fluxing station where the underside of the board is coated with a

Figure 3.18 Modern wave soldering unit
(Reproduced with permission of Technical Components Pty Ltd)

flux. The method of achieving this can either be by foam, wave or spray application. In the first, compressed air is bubbled through the liquid flux, causing it to foam upwards and coat the board. In the second, an impeller pumps a wave of liquid flux onto the underside of the board, and in the third, compressed air is used to finely spray liquid flux. In all three cases, viscosity or flux density, controlled by changing the amount of flux solvent present, is an important parameter in determining the amount of flux that coats the board. Correct flux application is important to ensure wetting and reduce bridges (Avramescu & Down 1986).

At the second station, boards pass over a preheat section. It serves several functions including reduction of thermal shock to the board when it enters the soldering station. Some board materials bow if heated on only one side, so preheat from the top as well as the bottom is often necessary. The preheat station also drives off the flux solvent because

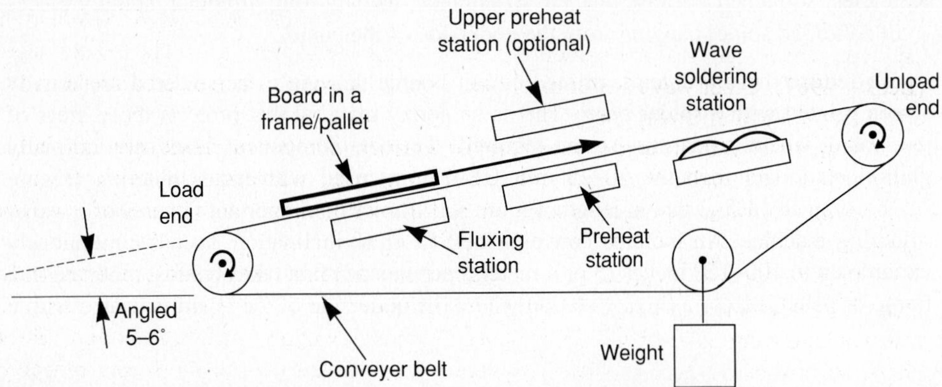

Figure 3.19 Wave soldering machine consisting of three stations

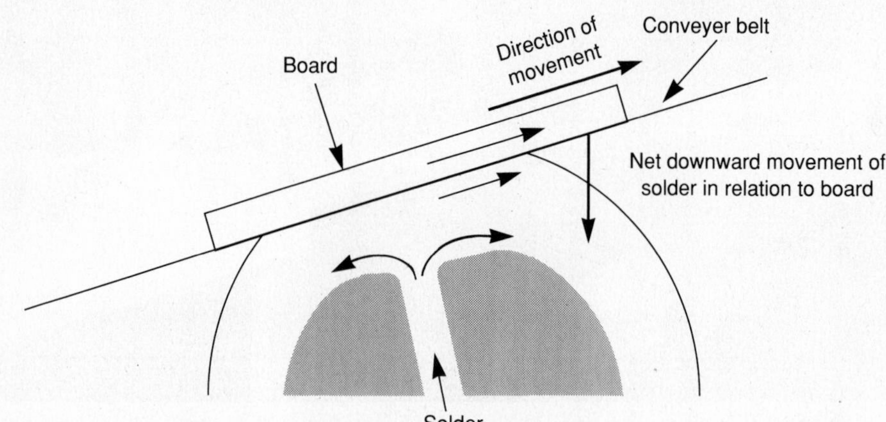

Figure 3.20 Important criteria when a board leaves a wave soldering machine

if this is still present when the board enters the soldering station, its evaporation can cause solder splatter. The preheat temperature is typically set in the range 90–110°C.

The final station is the soldering zone, where molten solder is pumped to form a laminar wave on the underside of the board. The exit of the board from the solder is critical and to reduce the number of icicles, the belt and solder horizontal speeds are made equal so that there is only a net downward movement. This is illustrated in Figure 3.20.

To reduce the defect rate (often 1 to 5 defects per 1000 joints) various wave soldering shapes and numbers are used. Names to describe them include bidirectional, omega and lambda shapes. Dual and occasionally triple waves are used. In the case of the dual wave, the first solder wave is made turbulent to ensure good wetting, while the second or finishing wave is of laminar flow and similar to that shown in Figure 3.20 (Lea 1988). When surface mount components are on the board, the non-wetting portions of the components cause shadows, resulting in poor wetting and high solder skip defects. The simplest solution is to pump the solder up over the normal wave height to create a pressure which deforms the solder wave, pressing the solder up to the solder joint (Becker 1987). Other wave solder machines vibrate the solder wave to achieve the same effect (Avremescu & Down 1986).

The temperature profile of a wave machine is not normally used to set up a unit, but it is interesting to compare this method of soldering with other reflow methods. Figure 3.21 shows such a profile. Preheat temperature and the turbulent nature of the solder waves can be seen.

Another necessary element of a wave soldering station is an air knife. Immediately after the solder wave, while the solder is still molten, an air stream forcibly removes solder where wetting has not been achieved. Thus non-wetted areas, bridges and icicles are all removed. Because of this, non-wetted areas of track are easy to see, reducing inspection times (Lambert 1987). The setting up of the knife is critical. It must be as close

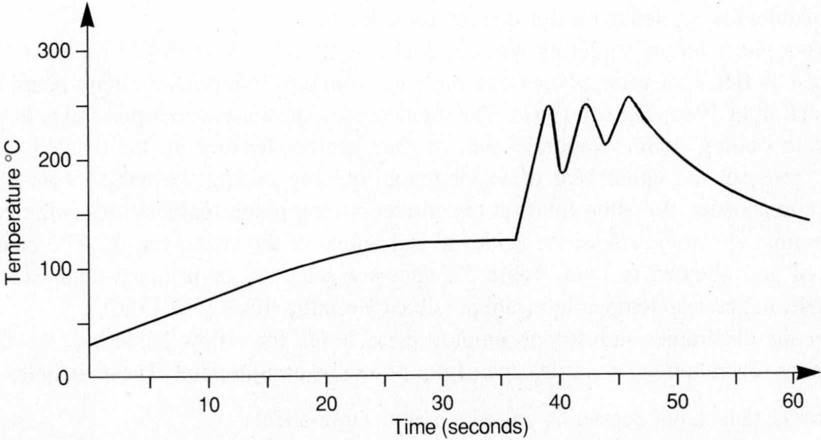

Figure 3.21 Time/temperature profile for a wave solder reflow machine

Table 3.4 Primary fluids available for vapor phase reflow soldering

Boiling point °C	Fluid trade names	Typical solders melt
174°C	Fluorinert FC43	70 Sn/18 Pb/12 In 100 In 58 Sn/42 In 58 Bi/42 Sn
215°C	Fluorinert FC70 Fluorinert FC5311 FC5312 Multifluor APF-215 Galden LS215	63 Sn/37 Pb 60 Sn/40 Pb 62 Sn/36 Pb/2 Ag 50 Pb/50 In
230°C	Galden LS230	
240°C	Galden HS240	
253°C	Fluorinert FC71 Galden LS215	100 Sn 96 Sn/4 Ag 60 Pb/40 Sn
260°C	Galden HS260	

as possible to the point where the board exits from the solder. The angle of impact of the air, air temperature and pressure are all important parameters that need to be correctly adjusted and sometimes changed to cope with varying board layout.

3.6 Vapor phase reflow soldering

The other three soldering processes are only suitable for reflow soldering. The components and boards must therefore be pre-tinned. Boards have extra solder paste added to all pads where reflow soldering is to occur. The methods used to apply this solder paste will be discussed in Chapter 6. It is sufficient to say that by using either pneumatic driven syringe dispensers or screen printing methods, the exact amount of paste required is applied to each pad ready for soldering.

Vapor phase reflow soldering was developed in 1975 by Western Electric (3M 1986) and used by Bell Laboratories for soldering wire wrap pins into printed circuit board back planes (Karpel 1986; Becker 1987). The method uses specially developed fluids that are heated to boiling point, evaporate and, as they are condensing on the printed circuit board, give out the latent heat of vaporization, quickly raising the temperature of the board components. By using fluids at the correct boiling point, temperatures sufficient to melt normal electronic solder are achieved and reflow of the solder occurs. The board is removed and allowed to cool. Table 3.4 shows a selection of primary fluids that are available. In between temperatures are possible by mixing fluids (3M 1986).

For the electronics industry to employ these fluids for reflow soldering, the fluids must have several other properties apart from a suitable boiling point. These include:

- inertness, that is, not degrading any of the other components
- high thermal and chemical stability so that they will not degrade with the long periods of boiling

- non-flammability and non-toxicity, so safe to use
- a dense vapor to minimize losses from any system.

Two types of machines are available at present. The most common is the batch machine illustrated in Figure 3.22. The second is the in-line machine used for large volume work which employs a conveyor belt to take the board into the primary vapor area. A second vapor area is not usually employed with the in-line machines; instead a closed loop vapor recovery system is used to return the air and vapor back to the tank (Pignato 1987).

In the case of the batch unit, a reservoir of liquid is heated to produce a zone of primary vapor at the boiling-point temperature. The upper limit of this zone is set by a ring of cooling coils. In order to minimize losses of the expensive primary fluid, a secondary inert vapor blanket is added and also contained by cooling coils. The materials commonly used for this blanket are based on trichlorotrifluoroethane, such as the refrigerant F113. Concern for the environment requires new products to be developed.

The operation of the unit is as follows. The cleaned boards are loaded onto the conveyor and lowered into the primary vapor through the secondary vapor area. Heating through condensation is almost immediate, typically taking 10–20 seconds. After a preset time the carrier and boards are moved to the secondary vapor area where they cool and it is ensured that all condensed primary liquid is returned to the chamber. After a further set time the now dry boards and carrier are removed from the unit.

The vapor phase method is certainly simple in concept and use. It allows mass soldering through fast and even heating to a well-defined temperature. It is not dependent on geometry or color of the components, there is no oxidization in the soldering zone, and it is tolerant to a wide range of fluxes. There is only one main variable to control—the time the unit is in the primary vapor.

Unfortunately, experience has shown that this presents some difficulties. Figure 3.23 shows a typical time/temperature profile for a vapor reflow solder unit. The rapid rise in

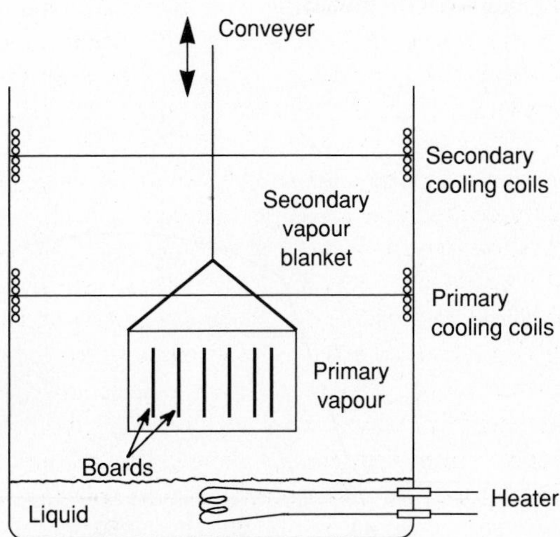

Figure 3.22 Batch vapor phase reflow soldering unit

temperature can cause two problems (Charbonneau 1986). Typically 50°C per second, it causes the solder paste to outgas, which can lift one end of a light component such as a small capacitor or resistor, causing what is called 'tombstoning' (if the component is raised near 90°) or 'drawbridge' or 'alligatoring' (if at some lesser angle). There is an open circuit. The solder can also form small solder balls through the outgassing. Varying the angle of dip of the board into the primary vapor by even a few degrees can help reduce this problem.

The second problem, due to the rapid temperature change, is the cracking of some components, particularly ceramic capacitors, due to the different thermal coefficients of expansion of materials used to make it. For example, barium titanate ceramic used as the dielectric has a temperature coefficient of expansion of 9–12 ppm/°C, the palladium silver electrodes approximately 14 ppm/°C, and the silver terminations on the outside approximately 18 ppm/°C. The thermal shock can cause the capacitor to fail. These problems can be overcome by preheating the boards or heat soaking them prior to reflow soldering. Temperatures range from 70–120°C and times from 30–120 minutes. Thus an additional plant item and extra time are needed.

Other minor difficulties are:

- The boiling points of the fluids change with atmospheric pressure or elevation above sea level. Thus the 215°C fluids only boil at 211°C if the factory is sited at an elevation of 3300 feet (1000 meters).
- Water is required for the cooling system.
- Slow thermal degradation of the secondary vapor does occur, forming small amounts of hydrochloric and hydrofluoric acids. Special plant must be added to remove these products.
- The primary vapor fluid must be filtered to remove foreign residue, principally flux residues.
- While the fluids are chemically inert and should not damage the ozone layer, there is some concern about long-term effects.

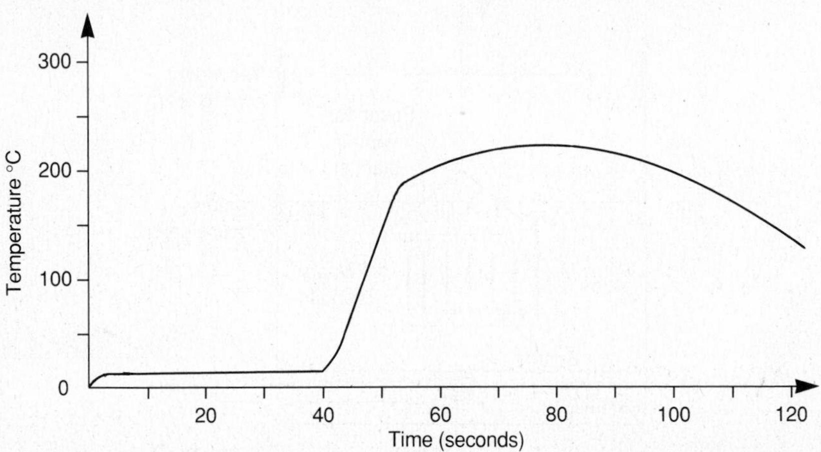

Figure 3.23 Typical time/temperature profile for a vapor phase reflow solder unit

3.7 Infrared reflow soldering

Belt furnaces have been used for thick film technology for several decades. The process requires only electrical power and, in some cases, an exhaust system for the solvents used. Unfortunately early attempts in transferring this technology from ceramic to laminated substrates ran into difficulties. When short wave lengths (< 2 microns) were used, heating was color dependent. Thus some components were excessively overheated and others insufficiently heated so both produced poor joints. Epoxy laminates are semi-transparent for frequencies in the range 0.75–3 microns (near infrared) and are absorptive in the 3–7 micron range (middle infrared) (Dow 1987). Recent infrared belt furnaces now use this middle infrared range and thus eliminate the color selectivity problem. Larger furnaces also employ hot air heating, with some 60% of heating achieved by convection and only 40% by infrared.

Figure 3.24 is a photograph of a small unit, the components of which are shown in Figure 3.25. Each of the zones has independent heating controls for upper and lower heating panels. Figure 3.25 also shows two typical time/temperature profiles. One has no preheat, while the other time/temperature profile sets the boards at 100°C for approximately one minute, to dry out pastes, drive off solvents, and heat the board. The first profile assumes that the solder pastes, with epoxy-mounted components, have been through an infrared epoxy curing furnace where preheating has occurred. The rate of rise of the temperature should not exceed 5°C/second, but for the profile with the preheat flat it will typically rise to the reflow zone at about half this rate.

Furnaces with as many as seven zones are now commercially available. These allow flexibility in setting up time/temperature profiles. Unfortunately, smaller infrared furnaces have little thermal mass and, for large board sizes, the profile can change considerably from one board type to another. Thus different control set positions are needed for boards of differing size and component density. In spite of these difficulties, infrared reflow

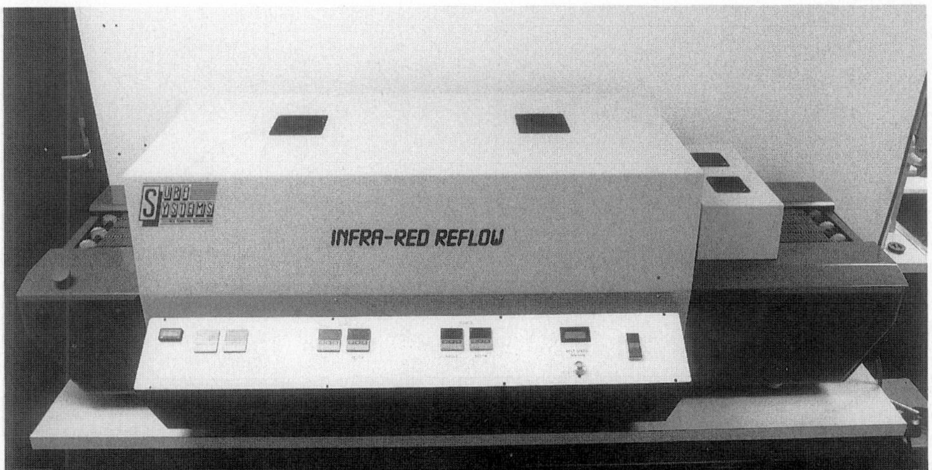

Figure 3.24 Small three-zone infrared furnace

soldering is growing in popularity because it provides an environmentally clean, low cost in-line facility.

3.8 Laser soldering

The use of lasers for soldering is a recent development. Although it has not been widely tested, the method has the potential for producing consistent, high quality joints quickly. The laser power is focused on each joint in turn and, while this may seem to involve a lengthy process, this is not the case. The time taken for reflow soldering of a joint can be but a few milliseconds, so that, with computer control, some 3000 joints per hour are possible.

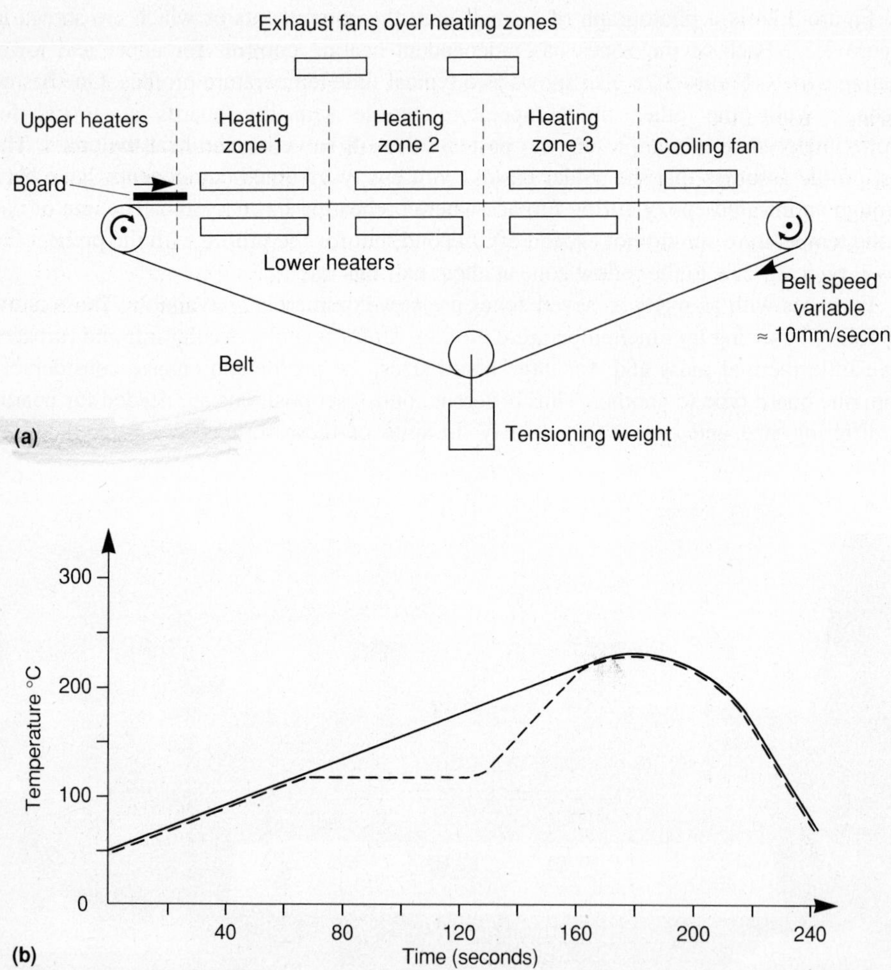

Figure 3.25 (a) Three-zone infrared furnace, and (b) time/temperature profiles for such a furnace

Two lasers are available: the carbon dioxide laser which emits energy at a wave length of 10.6 microns and the neodymium:yttrium-aluminum-garnet (Nd:YAG) laser at 1.06 microns. As shown in Section 3.7, epoxy boards respond differently to these two wavelengths, the first in the far infrared range, and the second in the near infrared range. When reflowing a joint, however, there are three components present: the solder, the flux, and the board. Each component responds differently to the laser beam (Lea 1987*a*).

At 10.6 microns wavelength the laminate absorbs the energy and becomes heated. Solder, on the other hand, has about a 75% reflectivity and therefore absorbs little of the energy, while the flux absorbs about 75%. Thus the heating of the joint is mainly due to the presence of the flux, transferring the heat to the solder by conduction. With the 1.06 micron wavelength, the solder only has a reflectivity of about 20% and absorbs the majority of the energy. The laminate has about the same reflectivity, but transmits the remainder of the energy rather than absorbing it. Thus, the heating effect of both lasers are of about the same order. There is, however, a distinct difference in the mechanisms. With the Nd:YAG laser of 1.06 micron wavelength, the beam does not have to be switched off as it transfers from one joint to the next and joints can be reflowed without having to use a flux.

The time taken to make a joint is as important as the power level. An 8 watt pulse 125 milliseconds long is a joule of energy and is sufficient to make an excellent joint. The same energy applied as a 100 watt pulse for 10 milliseconds will probably burn a hole through the pad (Lea 1987*a*). To control the amount of energy applied to a joint, an infrared detector is used to measure the joint thermal signature. The point at which the solder melts and flows can be seen from the infrared detector output and this information

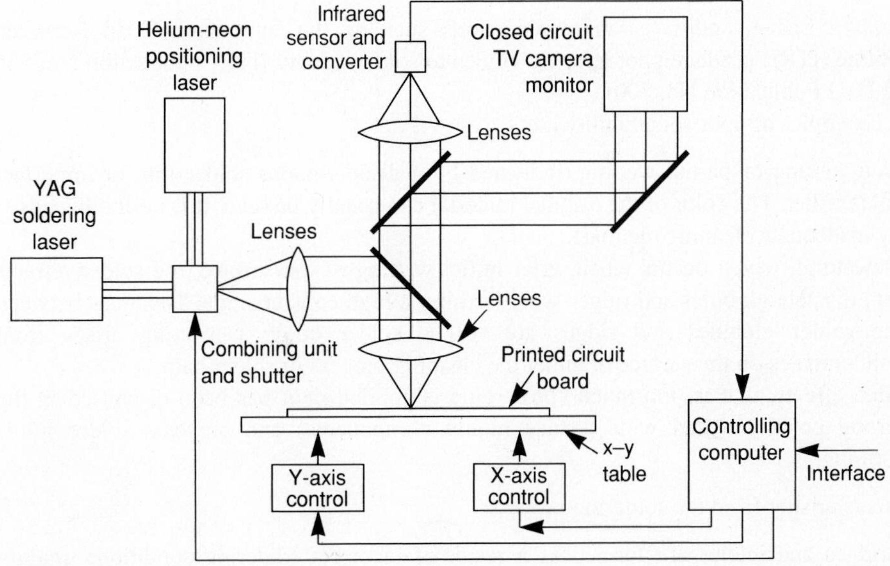

Figure 3.26 Laser soldering system with feedback to control the energy applied

is fed back to control the laser. Figure 3.26 shows such a system (*Electronics* 1986). A Nd:YAG laser is employed to do the soldering while the second helium-neon laser is used to position the spot accurately. Because carbon dioxide and YAG laser frequencies will damage human eyes, closed circuit TV is used to monitor the soldering progress. Typical laser output power is 12.5 watts with a minimum spot diameter of 0.6 mm. Such a system produces accurate, repeatable solder joints.

An interesting outcome of using the shorter wavelength is that the system can be used to solder joints on low melting point boards. Because the energy is restricted to a very small area and the board or laminate absorbs little energy, it is an ideal tool for soldering plastic molded printed circuit boards (Section 2.4). While the intermetallic layer formed by wave soldering is typically 1 micron thick, the thickness for a laser joint is estimated to be only 100 Å. This is because a laser joint can be made much faster. This may mean a less strong joint, but it will be more ductile.

Although this laser technique is the latest method of reflow soldering, it is estimated that already more than 5% of surface mount joints are made using lasers (Becker 1987).

3.9 Solder joint assessment

While there are various standard tests, solder joints on products are mainly assessed by visual inspection. Before any assessment can be made, the boards have to be thoroughly cleaned using chemical methods. Joints are basically classified into three categories:

1. preferred
2. acceptable
3. not acceptable.

Standard bodies and research organizations such as the International Tin Research Institute (ITRI) produce photographic guides to solder quality (ITRI Publication No. 555 and ITRI Publication No. 700).

Examples of poor solderability are:

- Non-wetting or partial wetting, indicated by a discontinuous solder film or imperfect solder fillet. The color of the original material can usually be seen. It is normally caused by inadequate cleaning methods.
- Dewetting, which occurs when, after initial wetting of the surface, the solder retracts into discrete globules and ridges which exhibit a high contact angle. The areas between the solder globules and ridges are still a solder color. Dewetting arises from contaminates on the surface of either the cleaning process or solder bath.
- Glue effects, that is, too much epoxy glue is applied or it has been dispensed in the wrong position (used with surface mount components) and prevents solder joints forming.

Defects arising from the soldering process:

- Bridges and icicles are formed as a result of incorrect soldering conditions, mainly when wave soldering is used.
- Blow holes or small spherical cavities arise from solidification of the solder around entrapped bubbles of air or flux (Lea 1987*b*).

- Degrees of joint filling. Solder fillets must be complete. In general, surface mount components will have smaller fillets than leaded components.
- Amount of solder in the joint. Too little solder will mean a joint of poor strength, while excess solder may mean not only unnecessary additional weight, but the excess solder can make it difficult to ascertain the joint quality. Excessive solder may be an indication of the solder bath temperature being too low.
- Misalignments. This is a surface mount problem, where components become misaligned with their pads during soldering. Poor design or differential heating will cause one end to melt before the other. Tombstoning, drawbridging or alligatoring may even occur.
- Shadows and skips. A problem when wave soldering is used with surface mount components. Incorrect orientation of components to the direction of travel of the board may cause this difficulty.
- Solder balling. There are two causes for this. First, flux being heated too rapidly can cause spluttering, and second, excessive oxide on solder particles does not allow them to coalesce before solidification occurs.
- Wicking and drainage. Wicking is caused by differential heating of leads of the component. It is common with J lead devices where heating of the upper portion of the lead draws the solder away from the pad area. Drainage is the reverse, where solder flows away from the pad into nearby vias and other component pads.

Table 3.5 summarizes some of the more common process-related solder joint defects and their causes (Elliot 1986).

Incorrect design of the board also causes defects. Most organizations have internal standards for board design which dictate optimum component termination size, minimum line width and spacing, etc. There are many other more subtle factors that need to be addressed, however, such as component spacing and orientation, and large copper areas on a board acting as heat sinks. Some of these difficulties will be discussed in Chapter 10.

3.10 Questions

1. Before hand soldering, why is it necessary to clean component leads and boards? What cleaning processes can be used?
2. You must recommend plant to your employer for reflow soldering surface mount boards. Name four processes that can be employed. Make a comparison chart for your employer.
3. Why does the electronics industry employ 60/40 solder? What is meant by this term?
4. Why is a flux used in making a solder joint? Can a flux be considered as a catalyst?
5. Explain the following terms:

   (a) dewetting
   (b) core solder
   (c) icicles
   (d) air knife
   (e) fluorinert.

Table 3.5 Some of the common process-related solder joint defects and their causes

Symptom	Solder temp. high	Solder temp. low	Solder wave height high	Solder wave height low	Solder wave uneven	Solder contaminated	Excessive solder dross	Preheat temp. high	Preheat temp. low	Flux contaminated	Flux spec. gravity low	Flux spec. gravity high	Flux no longer active	Flux not making contact	Flux foamhead low	Flux uneven	Flux blow-off excessive	No flux blow-off	Pallett too hot	Conveyor speed high	Conveyor speed low	Conveyor vibration	Conveyor angle high	Conveyor angle low	Early removal of board	Board not seated right
Insufficient solder flow through	√	√		√	√	√		√	√	√	√	√	√	√	√	√	√	√	√	√			√			√
Insufficient solder (solder side)	√			√	√	√		√	√	√	√	√	√	√	√	√	√		√	√			√			√
Dewetting or non-wetting		√		√		√			√	√	√	√	√	√	√	√	√	√	√	√	√					√
Solder voids or outgassing	√	√		√	√				√	√	√	√	√	√	√	√		√	√	√	√		√			√
Excessive solder (solder side)		√			√	√					√	√	√			√		√	√	√	√			√		√
Icicles		√		√	√	√		√	√	√	√		√	√	√	√	√		√	√		√		√	√	√
Bridging		√		√	√	√		√	√	√	√		√	√	√	√	√		√	√	√			√		
Webbing		√				√	√	√		√	√		√	√	√	√	√		√	√	√					
Solder balls and splatter	√		√		√				√	√	√							√		√						
Rough or disturbed solder		√		√	√	√	√													√		√			√	√
Grainy solder		√				√	√		√											√		√			√	
Cold solder joint	√	√				√			√											√	√	√			√	
Discolored solder joint	√					√		√	√	√										√	√					
Flux entrapment									√	√		√						√		√						
Blistering	√							√													√					
Measling	√							√													√					
Components lifted			√		√												√					√	√		√	√
Wrapage	√							√	√												√					√

6. Operations using molten metals and toxic gases require adequate safety precautions. What measures do you suggest for:

   (a) a wave soldering machine?
   (b) vapor reflow station?
   (c) hand soldering workstation?

7. Rigid and clinched solder joints are used in the industry. With the aid of drawings, define them. When would you use one in preference to the other?
8. With reference to Table 3.5, what corrective methods would you take if, after wave soldering, boards showed:

   (a) excessive solder?
   (b) dewetting?
   (c) solder balls and spatter?

3.11 References

Avramescu, S. & Down, W. H. (1986), 'Improved wave soldering', *Canadian Electronics Engineering*, **30** (November), pp. 32–6.

Becker, G. (1987), 'From soldering iron to laser: A review of soldering methods for surface mounting', *Hybrid Circuits*, **12** (January), pp. 22–7.

Charbonneau, R. A. (1986), 'Infrared vs vapor phase', *Circuit Manufacturing*, **26** (September), pp. 27–33.

Dow, S. (1987), 'An examination of convection/infrared surface mount reflow soldering—Part 2, *Surface Mount Technology*, **1** (April), pp. 13–23.

Dummer, G. W. A. (1978), *Electronic Inventions and Discoveries*, Pergamon Press, Oxford.

Electronics (1986), 'Intelligence comes to laser soldering', *Electronics*, **59** (10 July), pp. 75–7.

Elliott, D. A. (1986), 'Solder defect troubleshooter', *Circuits Manufacturing*, **26** (July), pp. 33–8.

ISHM (1984), *Surface Mount Technology*, International hybrid microelectronics, (Technical Monograph 6984–002), Silver Spring.

ITRI Publication No. 555, *Photographic Guide to Soldering Quality*, International Tin Research Institute, Uxbridge, UK.

ITRI Publication No. 700 (1988), *Soldering Surface Mount Devices*, International Tin Research Institute, Uxbridge, UK.

Karpel, S. (1986), *Tin and Its Uses*, No. 150, International Tin Research Institute, Uxbridge, UK, pp. 14–17.

Lambert, L. (1987), 'The air knife experiment', *Electronics Today International*, January, pp. 22–4.

Lea, C. (1987*a*), 'Laser soldering of surface mounted assemblies', *Hybrid Circuits*, **12** (January), pp. 36–42.

Lea, C. (1987*b*), *Blowholding in PTH Solder Fillets*, Wela Publications, Ayr, Scotland.

Lea, C. (1988), *A Scientific Guide to Surface Mount Technology, Electrochemical Publications*, Ayr, Scotland.

Manko, H. H. (1964), *Solders and Soldering*, McGraw-Hill, New York.

Pignato, J. (1987), 'New phase in vapor phase', *Circuits Manufacturing*, **27** (September), pp. 71–6.

RAAF (1972), *The RAAF Manual for High Reliability Hand Soldering*, Australian Air Publication, 7002.020-1, Department of Air, Canberra.

3M (1986), *Fluorinert Liquids Vapor Phase Condensation Heating Reference*, Industrial Chemical Products Division/3M, St. Paul, Minn.

4 Non-soldering joining methods

4.1 Introduction

Joining methods can be categorized under the headings of metallurgical, physical, mechanical, and chemical. Table 4.1 lists the major processes used in the electrical/ electronics industry. Soldering has already been discussed in Chapter 3 and brazing is seldom used in electronics because of the high temperature required (500°C). Microwelding can require components with special leads, normally containing nickel, and is therefore not popular. Thermocompression, ultrasonic and thermosonic methods are used for bonding out integrated circuits directly, minimizing the component footprint size and allowing a high packing density. The mechanical methods of crimp, wire wrap, and screw have important niche areas. For example, crimped co-axial connector joints are far more robust and reliable than soldered connections. Finally, the chemical method of conducting cements allows a joint to be made without applying any heat or force.

As with soldering, all the materials used in any joining process must be clean or else poor jointing may result.

Each of these joining methods, excluding soldering and brazing, will be discussed in this chapter.

4.2 Microwelding

Two types of microwelding are used in the assembly of cards: resistance spot and fusion welding (Zimmerman & Lewin 1983). Both allow the joining of two pieces of the same metal without using any extra metal or material. With resistance spot welding, electrodes clamp the component lead to the board and then a current I is passed for a set time t, generating energy E, given by:

$$E = I^2 R\, t \text{ joules} \qquad [4.1]$$

where R is the resistance of the contact.

Through the heat and pressure applied, fusion of the metals occur. This method was very popular in the 1960s for the assembly of densely packed cards using small outline packages. Figure 4.1 shows a typical system.

The electrodes are normally copper with conical tips, the diameter at the end being about five times the thickness of the flat leads being welded. In this particular case a simple capacitor discharge system is used to supply the energy. For consistent resistance

Table 4.1 Joining methods used in the electronics industry

Type / Factor	Solder	Braze	Weld	Thermo-compression	Ultrasonic	Crimp	Screw	Wrap	Conducting cement
Heat	Yes (270°C)	Yes (500°C)	Yes	Yes (300°C)	No	No	No	No	Slight (70–150°C)
Bond formed by	Wetting	Wetting and diffusion	Diffusion	Plastic flow and diffusion	Plastic flow and diffusion	Pressure and plastic flow	Pressure	Pressure and plastic flow	Chemical action
Process reversible	Yes	Yes	No	No	No	No	Yes	Yes	No
Ease of repair	Simple, slight degradation of components	Simple, but component damage	—	—	—	—	Simple, usually no damage	Simple, wire damage	—
Ease to automate	Easiest to do	Difficult and normally not done	Difficult and normally not done	Difficult and expensive but done	Difficult and expensive but done	Difficult	Not usually done	Difficult yet do under computer control	Not normally done
Joint	Good	Good	Good	Good (purple plague)	Good	Good	Good. Affected by vibration	Good	Good. Joints often mechanically weak and high resistance
Equipment complexity: manual and/ or automatic	Manual simple, wave soldering mod. complex	Manual simple	Moderately complex to do manually	Expensive and complex	Expensive and relatively complex	Manually simple. Automatic mod. complex	Simple	Manually simple. Automatic can be very expensive	Manually simple

Table 4.2 Examples of laser welding systems

Type	Mode of operation	Wavelength microns	Average power	Pulse length msec.	Pulse repetition frequency Hz
Nd:YG	Pulsed Continuous	1.06 1.06	1–400 1–1000	0.01–10 —	0.1–100 —
CO_2	Pulsed Continuous	10.6 10.6	1–400 50–5000	.05–.15 —	1–5000 —

welding, nickel is the preferred material. Consequently, component leads are made from nickel and the copper on the board is also nickel plated. These requirements have almost eliminated resistance spot welding as a means of interconnection.

With fusion welding, a metallurgical bond is formed through liquid-phase coalescence of the materials. The energy for coalescence can be gained from a variety of sources, including a TIG arc, electron beam or laser (the latter is becoming the most popular). As with soldering, both carbon dioxide and Nd:YAG systems are available. Table 4.2 summarizes some practical welding systems.

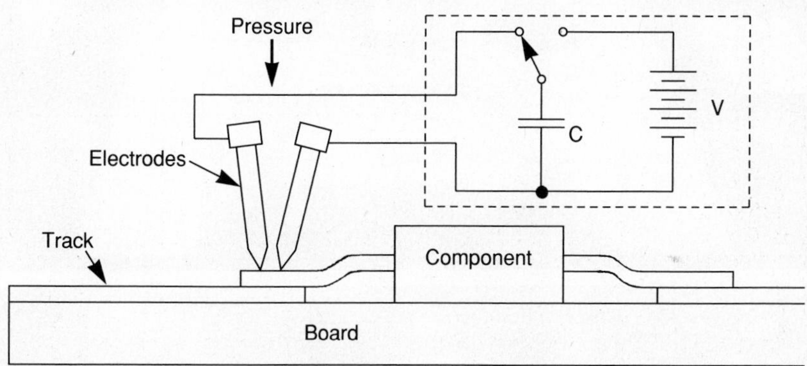

Figure 4.1 Simple resistance welder

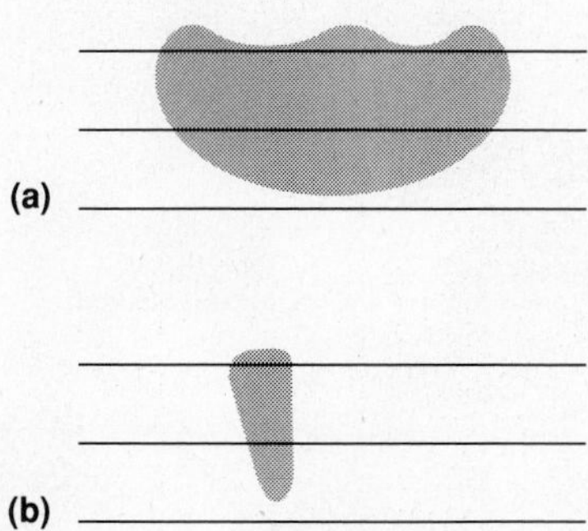

Figure 4.2 Cross section of: (a) a resistance weld, and (b) laser fusion weld

Laser welding systems give a deep penetration weld when compared to that from a resistance weld. This is illustrated in Figure 4.2.

A difficulty with laser fusion welding (it equally applies to laser soldering) is the high reflectivity of the metals being welded. The reflected energy causes degradation to other packages on the board, shown as surface carbonization (Lea 1987).

Figure 4.3 Ultrasonic bonder for chip-and-wire bond assembly

Welding methods are employed for mounting TAB and other special integrated circuit packages.

4.3 Wire bonding

In card assembly, wire bonding is normally restricted to the direct bonding out of an unencapsulated semiconductor die and is called the chip-and-wire bond process. The die is mounted to the board, in the correct orientation, using epoxy (silver loaded if an electrical connection is needed) or a molybdenum gold tab eutectically bonded to the die and then soldered to the board. Three types of bonding methods are used:

1. thermocompression
2. ultrasonics
3. thermosonic.

Thermocompression requires that the substrate be heated to 300–350°C, which is excessively high for most laminated boards. Consequently, it is seldom used. Ultrasonic bonding is undertaken without any heating, the energy to form the joint being supplied from a 40 kHz generator through a magnetostrictive transducer and focused on the joint area with a tapered horn and wedge. Pressure and the ultrasonic vibration cause plastic flow of the materials to form the joint. Figure 4.3 shows an ultrasonic wire bonder.

The third method of thermosonic bonding is a combination of ultrasonic and thermocompression. The substrate is heated to 150–180°C and the bond is made ultrasonically. The process is more tolerant to material variations than straight ultrasonic bonding. Figure 4.4 shows a chip that has been wire bonded direct to the board.

Figure 4.4 Silicon chips wire bonded direct to a laminated board

Both bonding methods are suitable for attaching either aluminum or gold wires to soft gold surfaces. Unfortunately, the gold used in the printed circuit industry is hardened to improve its life on connector fingers. Wire bonding to this type of gold, even with thermosonic methods, can be unreliable. Bonding to freshly cleaned copper is readily achieved, but the presence of any oxide will degrade and possibly prevent bonding occurring. The solution is therefore to apply a gold flash, typically 0.2 micron thick to the cleaned copper surface so that, on bonding, the gold flash is removed by the process, exposing the clean copper surface underneath. In some cases, a 1 micron thick plated layer of nickel is added to the copper surface to improve the quality of the bond. It is still covered by the gold flash.

Bond strength may be tested with a pull tester and should not fail with pull forces less than 5 g for 25 micron gold wire or 1 g for the same diameter aluminum wire.

4.4 Wire-wrap

The wire-wrap process was developed by Bell Telephone Laboratories for the telephone industry in the United States. They required a joining method that did not need heat, that could be undertaken by people with few skills, was reversible, could survive at least 50 makes and unmakes, and have a life of some 40 years.

As shown in Figure 4.5, the joint consists of a solid wire wrapped the correct number of turns around a post. The post must have the correct radius of curvature and the wire the correct elongation coefficient so that, at the corners of the wrap, the post bites into the wire to form a gas tight, high-pressure joint, where solid state diffusion of the metals has occurred.

In the original standard wire-wrap joint, the insulation on the wire was only wrapped half a turn around the post. It was found that, where there was movement and vibration, the wire fractured at the point where the insulation ends. To prevent this, a modified wire-

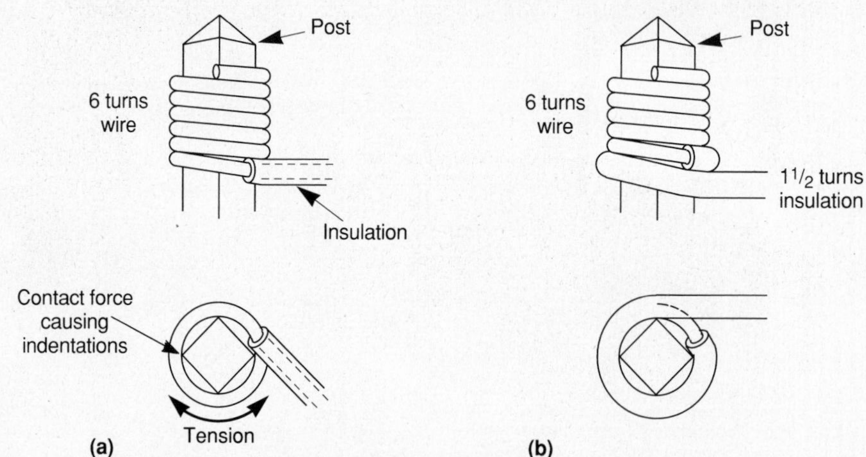

Figure 4.5 (a) Standard and (b) modified wire-wrap connections

Table 4.3 Recommended number of turns for wire-wrapped joints using different wire gauges

Wire diameter		Wire gauge AWG	No turns
inches	mm		
0.0113–0.0126	0.28 –0.315	28–29	12
0.0179–0.0226	0.455–0.574	23–25	6
0.0253–0.0359	0.643–0.920	19–22	5

wrap joint is now employed where one and a half turns of insulation are used. Further, in the original system, posts were placed on 0.2 inch (5.08 mm) spacing, but for application to the printed circuit board industry, where the standard dual in line package spacing is 0.1 inch (2.54 mm), this reduced spacing is now the standard.

The posts are normally square in cross section, although rectangular ones require a greater stripping force—that is, the force needed to pull the wrapped wire off the post. They are made from a range of materials, such as beryllium copper, phosphor bronze, brass or steel. The hardness of the post also determines the stripping force. For 24 AWG (0.511 mm) copper wire, the stripping force may be in the order of three kilograms. The posts can be plated with gold, silver, or more usually tin lead. The length of the posts determines the number of connections that can be made to it and is typically in the range 3–5.

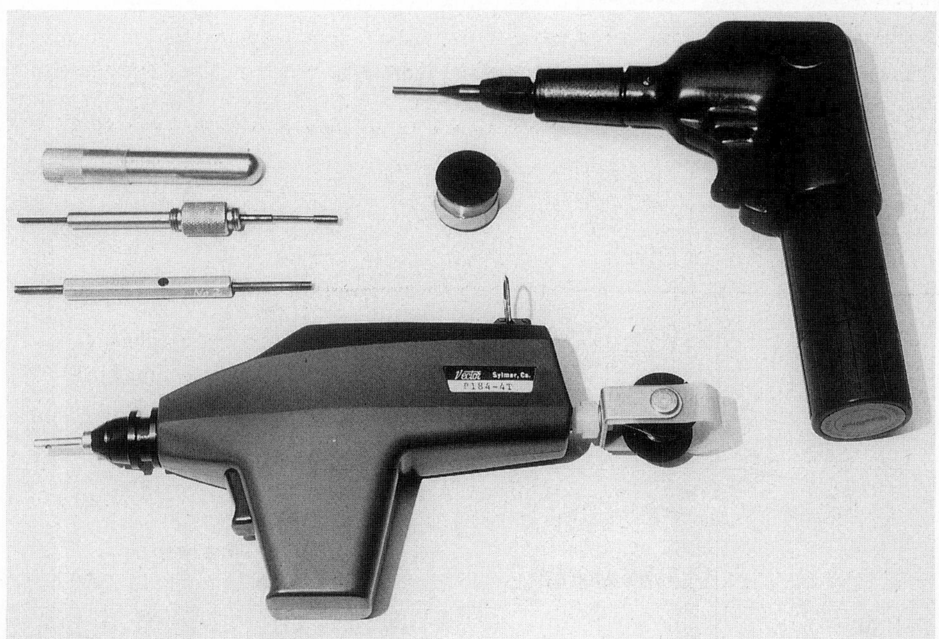

Figure 4.6 Manual wire-wrap tools

The wire used in this joining method is solid and must be capable of 15–20% elongation, to maintain pressure around the post. Tin-plated copper wire is commonly used, but brass, kovar and other materials are sometimes employed. The number of turns wrapped around a post is dependent upon the gauge of the wire. Table 4.3 shows recommended turns for a range of wire sizes. Wires of 24 AWG and 28 AWG diameter are the most common.

Stranded wire can be wrapped around a post, but must be soldered for permanency.

The preferred insulation for the wire is polyvinylidine fluoride (Kynar) as it has extremely good abrasive and cut-through resistance, necessary attributes for wires being fed through a maze of posts. PVC, nylon and Teflon are used to a lesser extent.

Tools available for making wire-wrap joints can be manual, semi-manual or fully automatic. Hand tools often have an insulation stripper built into them and a slit in the side to the right depth so that, on wrapping, the correct number of wire and insulation turns is achieved. Figure 4.6 shows several manual tools and an electric wire-wrap gun.

One tool has a spool of wire attached to it. No stripping of the insulation is required because, on wrapping, the post corners cut through the insulation. This method is only recommended for experimental prototyping. The simple tools have a wrapping tool one end and an unwrapping tool at the other. Joints can be made and unmade in under two seconds. With training and good quality control, joint reliability can be superior to soldered joints by a factor of 20. Typically, wire-wrapped joints can be 0.005 FITS (or Failure unITS and is the number of failures in 10^9 component hours. See Section 5.6). The quality of the joints can be assessed by:

- inspection
- performing stripping force tests
- performing unwrap tests.

The last two are destructive tests, illustrated in Figure 4.7.

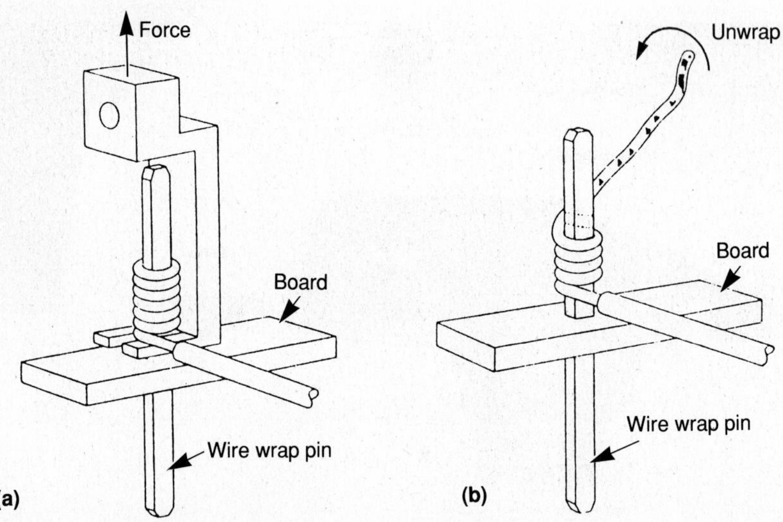

Figure 4.7 Stripping: (a) force and (b) unwrap test procedures

When inspecting, one should look for consistency of wrap, ensure there are no gaps between turns, no excessive tail ends, no overlapping of turns, no damaged insulation, a small spacing between different wrapped wires, and the correct number of turns for both wire and insulation.

In the stripping force test, the unit pulls along the length of the post and measures the force required to remove the wire. In the unwrap test, the wire is unwound and should not break. If it does, the post corners have cut too deeply into the wire—that is, the wrapping was incorrectly undertaken or the post and wire materials were not compatible.

In summary, the wire-wrap joining method has the advantages of:

- high density wiring capability
- automated wiring
- high reliability
- repairable joints
- no heat required.

However, it has the disadvantages of:

- special types of solid wires
- special wire wrap posts
- being mainly limited to low frequency and digital systems
- high density wiring is difficult to do manually because two adjacent posts on 0.1 inch (2.54 mm) spacing in a maze of pins can easily be confused. There is a high level of concentration needed, causing strain on the operator's eyes.

Figure 4.8 shows a typical wire-wrap board where all dual in-line packages have been mounted in wire-wrap sockets.

Figure 4.8 Wire-wrap digital board

4.5 Crimped joint

This is a method of joining where compression is used to produce plastic flow, resulting in a uniform, gas tight, metal-to-metal bond. Both wire and insulation are crimped to give a very robust connection. There are numerous styles of crimp joints, including radio frequency connectors. In the assembly of electronic cards, crimping is often used to make interconnections from the board to some other part of the system. Ring, spade, pigtail, and other terminal styles are used. Figure 4.9 shows a selection of styles.

The wire used in the connection is not critical. Both copper and aluminum can be employed, either solid or stranded, the latter being most common. Terminals are also available in a range of materials, including brass, copper, nickel, stainless steel, aluminum, beryllium copper, and bronze. Care must be taken not to crimp dissimilar metals in equipment that must operate in humid conditions since galvanic corrosion can impair the long-term reliability of the joint. With some terminals the wire must have the insulation stripped off for a set length while, for others, the crimping action thrusts a spike through the insulation to make the electrical connection. The pressure exerted by the crimping tool (either manual or assisted) is critical, as shown in Figure 4.10. Too little pressure produces an electrically intermittent joint and too much pressure produces a joint that is mechanically weak. Most manual crimping tools are restricted to a particular terminal/wire size and have either a simple stop to ensure excessive pressure is not applied or a stop with a rachet operation so that the crimped terminal cannot be removed until the correct pressure has been applied.

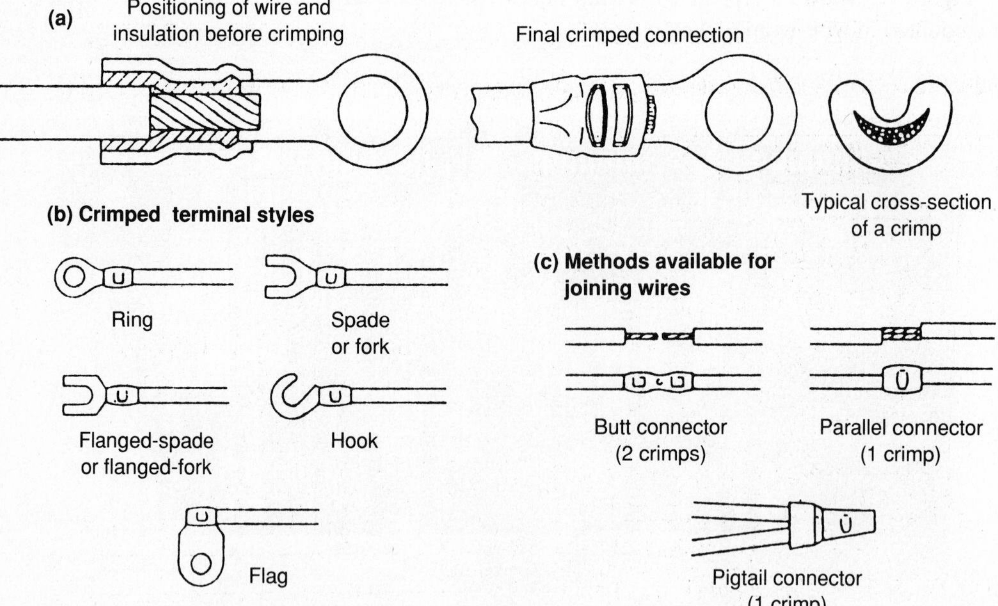

Figure 4.9 A selection of crimped terminal styles

The reliability of crimped joints is extremely high. This is fortunate as they cannot be repaired. Joints can be tested by either inspection or tensile strength tests. The latter is a destructive test in which the force is measured to break the wire or crimped joint. When inspecting a joint, points to watch for are correct wire and terminal sizes, correct indentation, wire strands protruding past the crimped area, and whether the insulation is correctly positioned and crimped.

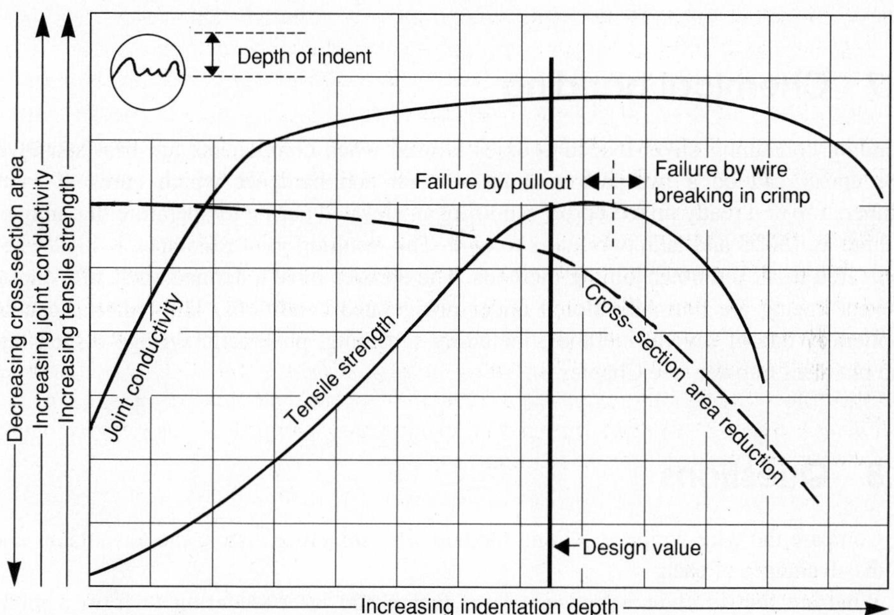

Figure 4.10 Variations in crimp termination characteristics with indentation depth (Harper 1970)

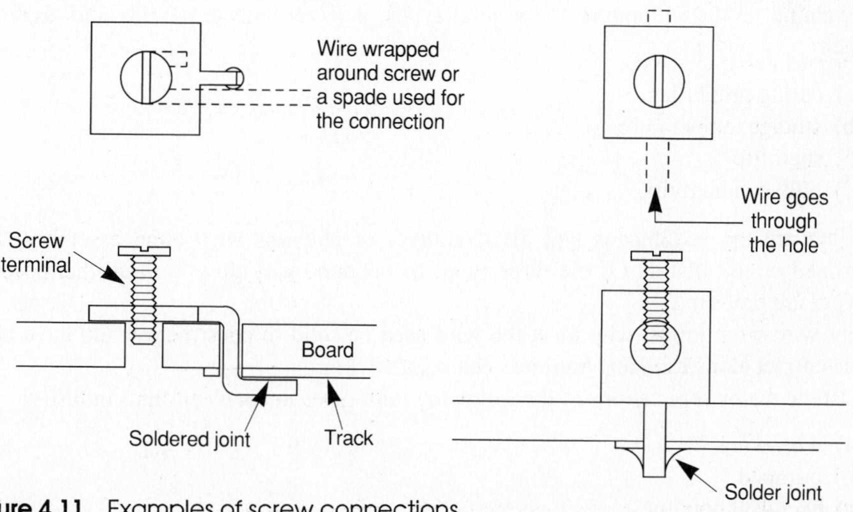

Figure 4.11 Examples of screw connections

4.6 Screw joint

Although the oldest of the electrical joining methods, the screw joint is still employed on laminated boards, mainly in the electrical and automotive industries. Star and other slip-proof washers are sometimes added to prevent the screw from turning. The two main versions used are illustrated in Figure 4.11. The first is frequently used with the crimped spade connection shown in Figure 4.9b.

4.7 Chemical bonding

Bonding containing silver-loaded epoxies is used when components are heat sensitive. The epoxy can be a two-part mixture of resin and hardener which cures at room temperature or a ready mixed epoxy requiring an elevated curing temperature that may be as high as 150°C and take two hours to cure. The resulting joint resistance is high when compared to all the other joining methods. The epoxies have a defined shelf life, and to prevent ageing are normally stored under refrigerated conditions. The material can be applied by one of several methods, including screening, pneumatic syringe dispensing, and pin mass transfer (see Chapter 6).

4.8 Questions

1. Compare the wire-wrapped joining method with soldering, listing the advantages and disadvantages of each.
2. What are the differences between laser fusion and laser soldering to form a joint? Which do you believe has the best long-term use in the electronics industry?
3. You are inspecting the quality of crimped joints. What would you look for? Are there any objective measurements (destructive or non-destructive) you would make?
4. Examine product literature for several types of silver-loaded epoxies and ascertain their:

   (a) curing properties
   (b) storage temperature
   (c) shelf life
   (d) bulk conductivity.

5. What are the advantages and disadvantages of chip-and-wire bond assembly to a printed circuit board? Of the three types of bonders, why do you think thermosonic types are preferred?
6. For wire-wrap joints, why must the wire used be solid in construction and have both the correct elongation and hardness characteristics?
7. List the major applications of the following joint types in the electronics industry:

   (a) wire-wrap
   (b) crimped
   (c) chemical bonding.

4.9 References

Harper, C. A. (ed.) (1970), *Handbook of Materials and Processes for Electronics*, McGraw-Hill, New York.

Lea, C. (1987), 'Laser soldering of surface mounted assemblies', *Hybrid Circuits*, **12** (January), pp. 36–42.

Zimmerman, D. D. & Lewin, D. H. (eds) (1983), *The Fundamentals of Microjoining Processes*, The Welding Institute, Cambridge (also issued as ISHM Technical Monograph Series 6983-003).

5 Electronic components

5.1 Introduction

How do you define electronic components? In the narrow sense they are the resistors, capacitors, inductors, and semiconductor devices that are assembled onto the card. However, in many instances, this is too narrow a definition. Consequently, in this chapter a broader approach will be used. In reality, the card onto which all other components are assembled is a component in its own right. For multilayer boards, it is probably the most expensive component in the module. The broader definition must include any element that is assembled with others to make up the card or module. Therefore, in addition to the resistors, capacitors, inductors, and semiconductors already mentioned, it includes connectors, test pins, heat sinks, switches, and wires. In a single chapter it is not possible to cover every possible component so it will deal only with those considered most important.

Conventional passive components (resistors, inductors, and capacitors), diodes, transistors and integrated circuits are normally produced in a range of standard packages that may be leaded or unleaded. The leaded varieties frequently come in two varieties, through-hole or surface mounting. Therefore, in discussing components and their performance, consideration will be given not only to component materials and construction but also to their method of packaging.

5.2 Passive components

Components normally have what is referred to as a life-cycle curve, which has three sections, as shown in Figure 5.1. New components normally undergo a period of experimentation before they become accepted throughout the industry. It takes time to build up a record of reliability and for designers to appreciate any advantages. Once this has happened they become a mature component in the industry and they are accepted and used. Finally, because it is a dynamic situation, newer components eventually supersede earlier ones and their use declines. Industries producing consumer products tend to accept newer components more readily than the professional and military services. For the latter it is important that long life is assured and therefore more time is required to build up confidence. Examples of components in the three stages of development are given in Figure 5.1.

Table 5.1 Preferred tolerance range for components

Tolerance									
10%	**5%**	**2% (every other value)**					**1%**		
10	10	100	133	178	237	316	422	562	750
	11								
12	12	102	137	182	243	324	432	576	768
	13								
15	15	105	140	187	249	332	442	590	787
	16								
18	18	107	143	191	255	340	453	604	806
	20								
22	22	110	147	196	261	348	464	619	825
	24								
27	27	113	150	200	267	357	475	634	845
	30								
33	33	115	154	205	274	365	487	649	866
	36								
39	39	118	158	210	280	374	499	665	887
	43								
47	47	121	162	215	287	383	511	681	909
	51								
56	56	124	165	221	294	392	523	698	931
	62								
68	68	127	169	226	301	402	536	715	953
	75								
82	82	130	174	232	309	412	549	732	976

Since no components are ideal, there are, in general, three areas that are important when considering components. They are:

1. the range available
2. the characteristics of the device and how they change with ambient conditions, circuit operation and age
3. the device reliability figures.

Under the first of these, there are three further important considerations. First, the range in terms of electrical parameters. A capacitor type may only be produced in the range 0.001 to 1 μF or a resistor from 10 ohms to 1 Mohm. If the circuit requires a greater range, then these types may not be entirely suitable. Second, it is impracticable to produce discrete components covering all possible values. Thus, components come in preferred ranges. Table 5.1 shows the tolerance ranges, with 5% being the most commonly used today.

The values are spaced so that they allow for the spread of component values during manufacture. Take, for example, a 3.3 kΩ resistor in the ± 10% range. Its value can be from 2.97 to 3.63 kΩ. Its lower limit matches the upper 10% limit of the lower 2.7 kΩ resistor (being 2.97 kΩ) and the lower limit of the higher 3.9 kΩ resistor (3.51 kΩ) slightly overlaps its upper limit.

This tolerance is normally the initial manufacturing tolerance and does not include other changes during component life. Some resistors are specifically made and marked to indicate that they will stay within the tolerance all of their lives. Components are often marked to indicate their value. In some cases this is a printing of a numerical value. Several manufacturers replace the decimal point by the unit to avoid confusion. An example is a 3.3 kΩ resistor shown as 3k3. Other manufacturers give three digits, the first two being the value and the last the scaling factor. Using the same resistor, it would be shown as 332. For very small components a color coding system is often employed as shown in Figure 5.2 and Table 5.2.

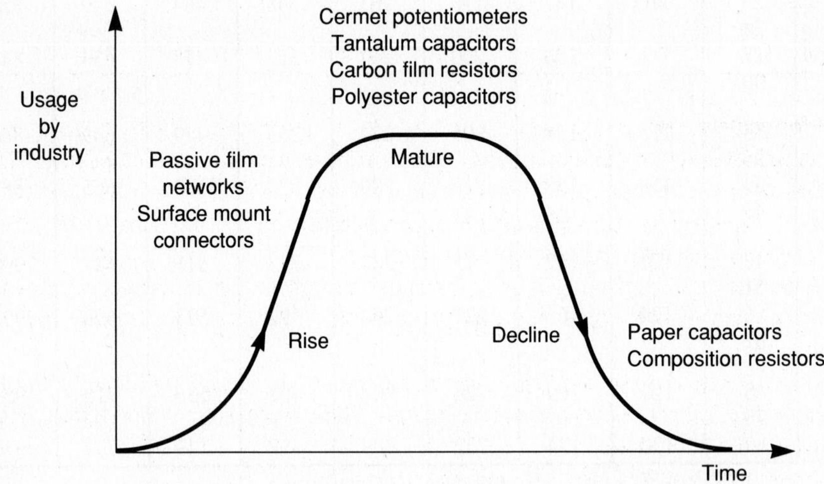

Figure 5.1 The life cycle of a component

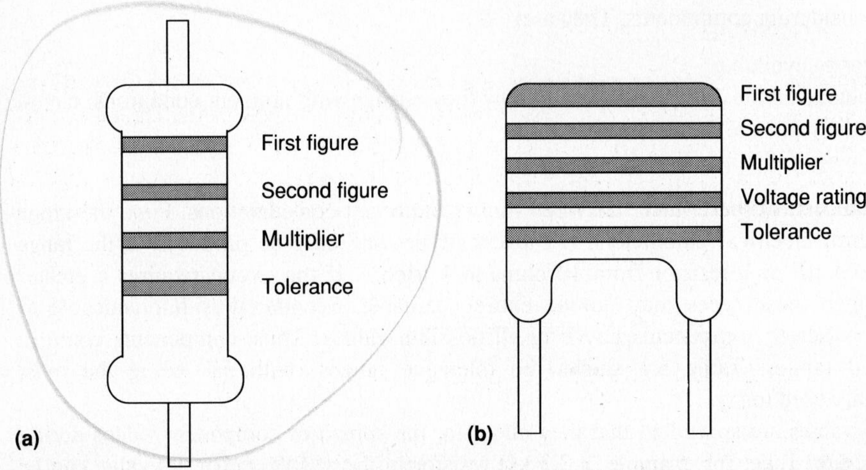

Figure 5.2 Color coding of: (a) resistors and (b) capacitors

Table 5.2 Color code interpretation

Color	1st and 2nd figure	Multiplier	Tolerance		Capacitor rated voltage
			Resistor %	Capacitor %	
Silver	—	0.01	10	—	—
Gold	—	0.1	5	—	—
Black	0	1	—	20	—
Brown	1	10	1	—	100
Red	2	100	2	—	250
Orange	3	1K	—	—	—
Yellow	4	10K	—	—	400
Green	5	100K	—	—	—
Blue	6	1M	—	—	630
Violet	7	10M	—	—	—
Gray	8	—	—	—	—
White	9	—	—	10	—

Unfortunately, no system is universal. With extremely small surface mount components, the actual component may have no marking on it at all, its value simply stamped onto the packaging, be it a plastic bag or tape.

The third factor under the heading of range available is the number of different package types. This topic is covered in Section 5.2.4.

The characteristics of devices and how they vary according to factors such as time, temperature and power are best considered under the heading of each individual component type.

Finally, the reliability of the device is an important aspect, particularly for large systems where failure of a component can cause difficulties, even emergencies. Section 5.6 is devoted to this topic.

Several component types will be discussed now in more detail.

5.2.1 Resistors

Resistors can be manufactured using a range of processes and materials. This includes carbon composition and film, wire wound using nichrome and other wire materials, metal oxides such as tin oxide, metal film such as nickel chromium alloys, bulk metal and thick film cermet pastes. The reason for such a large range of materials is simply a trade-off between cost and particular performance characteristics, be it low noise, high stability or small size.

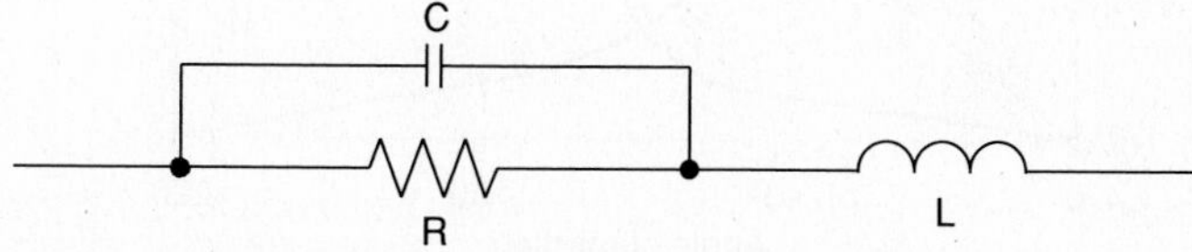

Figure 5.3 Lumped model of a resistor. C = 0.1–2 pF. L ≈ 0.1 μH if a leaded component

The physical construction of a resistor is often a ceramic or glass cylinder or rectangle onto which the resistive material is deposited. End caps are used to make the connection, forming a physical pressure contact with the resistive material (see Figure 5.12). As previously stated, no component is ideal and resistors are no exception. There is a shunt distributed capacity, lead wire and bulk inductances so that it can be represented by a simple lumped model as shown in Figure 5.3.

Consequently, the impedance of a resistor is frequency dependent. The larger the value of resistance, the lower the frequency at which the reactive components start to dominate. While manufacturer's data should be consulted, small fixed film resistors (< 500 mW), approaching 20 kΩ in value can be considered ideal to typically 100 MHz and resistors less than 1 kΩ to 300 MHz.

Most resistors are of a fixed value because it is good design practice to minimize, and preferably eliminate, variable components. Not only are they more expensive, but the greater the number of variables, the more difficult it is to set up a card.

With variable resistors, there is normally a screwdriver adjustment so that the value can be set on the card, perhaps to zero the offset of an operational amplifier. Variable resistors are either carbon film, wire wound or planar thick film printed form, each with a metal wiper to make the variable contact. When all three connections are used, the unit is called a 'potentiometer'. The resistance with movement of the wiper can follow a number of laws from antilog through linear to logarithmic. These responses are illustrated in Figure 5.4. A letter or sequence of letters is often used to specify this property. Thus a linear potentiometer may be indicated by the letter A or LIN.

There is a whole range of temperature effects that may be significant. First, a resistor has a maximum dissipation figure. There is normally a linear derating with ambient temperature, as shown in Figure 5.5.

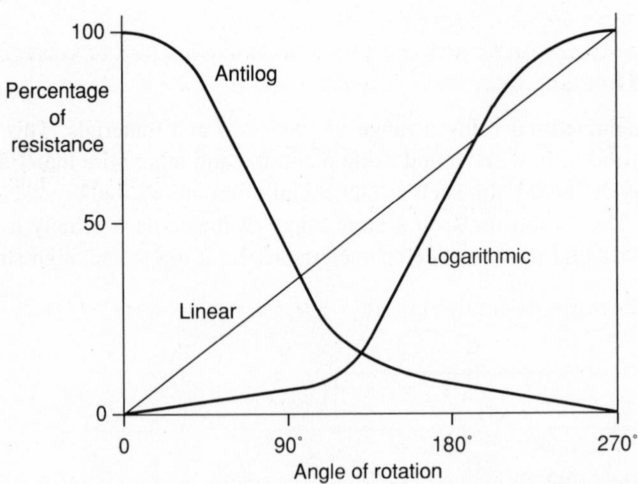

Figure 5.4 Examples of resistance versus angle of rotation for the potentiometer

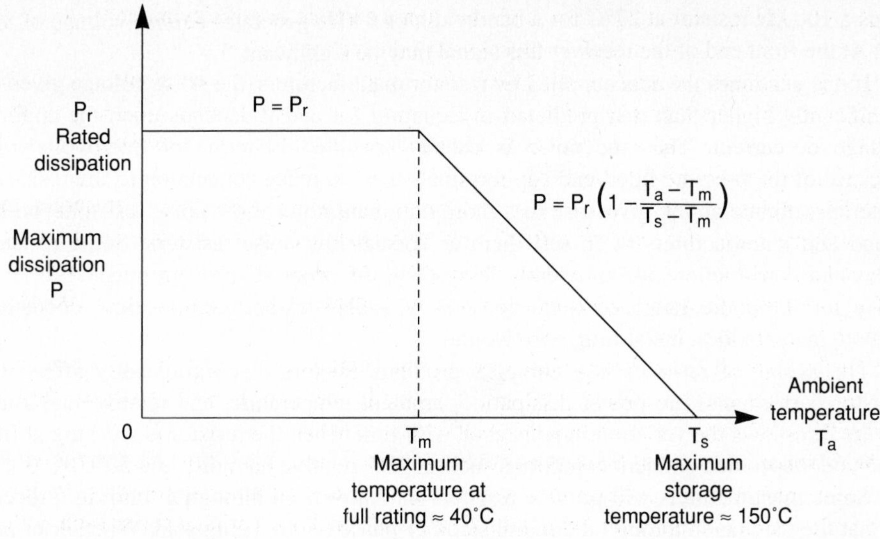

Figure 5.5 Dissipation characteristics of a resistor

Wire wound power resistors should always be run near their maximum power rating to ensure that all water vapor is driven off. This prevents corrosion causing the component to become open circuit.

Resistors also have a temperature coefficient of resistance. It is normally small in a good quality resistor, negative in value for carbon material and positive for others. When composition carbon and wire wound resistors are operated near their dissipation limit (from constant voltage/current sources respectively), then the carbon resistors tend to short circuit fail, while the wire wound type will go open circuit.

Finally, care must be taken when soldering a resistor into a circuit. If not, the temperature can exceed the maximum storage temperature for too long, causing a permanent offset to the value of the component amounting to 1 or 2%.

Resistors have a voltage rating and voltage coefficient and while, for normal voltage levels, these are insignificant, there are occasions when both have to be considered. Most resistors will work up to several hundred volts. With some types the DC and AC voltage limits may not be the same. Voltage transients above the rated value may also induce permanent changes in resistance value. An example where the voltage coefficient should be considered is in the case of precision attenuators working at even moderate voltages. To eliminate the effects of voltage coefficients, all resistors in the attenuators should be of the same material.

Resistors generate white or Johnson noise which can be described by the relationship below:

$$e = \sqrt{5.5.\ 10^{-11}\ T.R.\ \Delta F} \qquad [5.1]$$

where e is the RMS noise voltage in microvolts
T the temperature in degrees kelvin
ΔF the bandwidth in hertz
R the resistance in ohms

Thus a 100 kΩ resistor at 27°C for a bandwidth of 5 kHz generates a noise voltage of 8.3 μV. At the front end of the receiver this signal may be significant.

If one examines the data supplied by resistor manufacturers, the noise voltage given is significantly higher than that predicted in Equation 5.1 and it depends upon the applied voltage or current. Thus the noise is usually specified in terms of microvolts/volt. Because of the pressure fitted end cap terminals used to make connection to the resistive materials, minute arcing gives rise to a more dominant noise. Low noise resistors can be found and manufacturers often sell them as special low noise resistors. Some resistor types and constructions are inherently less noisy, the progression from noisiest to least noisy for the audio range of frequencies is typically—carbon composition, deposited carbon, metal oxides, metal film, wire wound.

The ageing of resistors is a complex problem. Factors that significantly affect the resistor performance are power dissipation, ambient temperature, and relative humidity. Figure 5.6 shows the variation in resistance with time when the resistor is working at full load under specified ambient conditions such as 95% relative humidity and 50°C.

Some manufacturers will provide resistors that have been through a 'burn in' process so that they are taken through the initial stability phase before being sold. Whether or not this is critical depends upon the circuit in which the component is operating and its sensitivity to component variation. While a 1% change may seem small, it may only be one of several changes which, collectively, could be significant. For example, a company manufacturing equipment for the armed services and using purchased 5% resistors in the 10% range, used 20% tolerance in their design, arrived at as follows:

Initial tolerance	± 5%
Soldering in circuit	± 3%
Temperature cycling –55 to +85°C	± 2%
Humidity and loading cycle	± 10%
Total	**± 20%**

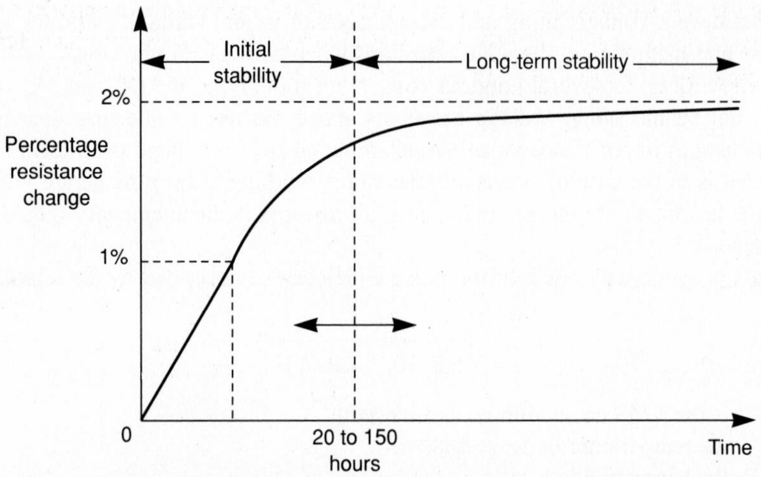

Figure 5.6 Variation in resistance value with age

The three approaches taken to design circuits to ensure that they are tolerant to component variations are:

1. worst case
2. statistical
3. Monte Carlo.

The first is the traditional method because it simplifies calculations. In effect, a rectangular distribution of component values is assumed. At times the 10 and 90 percentile points are used. It builds in a large safety factor, but can give rise to unrealistic values (such as a negative resistance) and circuits that consume more power than necessary. With the growth of computers, the latter two methods are now popular. Gaussian or other distributions are assumed for the statistical approach and results in near optimum component values. The difficulty with this method is knowing whether you are using the correct distribution curve, something component manufacturers never reveal. You would be correct in assuming that a stable process would have a Gaussian distribution, but what you do not know is the post processing that the manufacturer has undertaken, be it component selection and batching, burn in, etc.

In the final design method, the designer employs mean values and then evaluates the design by simulating a production run, using Monte Carlo methods. If the yield is sufficiently high, the design is acceptable.

Computer sensitivity analysis of component tolerance is used in all three approaches to verify which are the critical components and, for these, more stable types may be inserted.

Table 5.3 summarizes many of the resistor types that are available today.

5.2.2 Capacitors

Capacitors, like resistors, are also non-ideal and have a lumped model which is of the same configuration as resistors, except that the capacitor is the desired component and the resistor represents a loss. A modified serial lumped model is also important for some capacitor types (see Figure 5.7), such as aluminum oxide electrolytic capacitors.

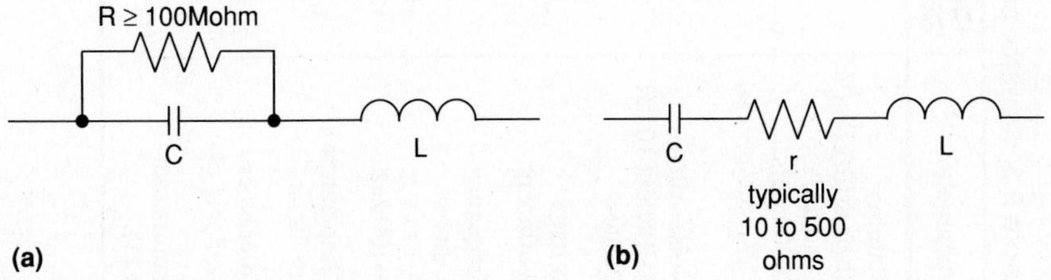

Figure 5.7 Small signal lumped models of capacitors: (a) for most types, and (b) for some electrolytic capacitors

Table 5.3 Resistor comparison chart

Parameter	Carbon composition	Carbon film	Wire wound	Metal oxide	Metal film	Bulk metal	Metal glaze (thick film)
Stability: Shelf drift/year (%)	4	0.75		0.1	0.03	0.0025	0.1
Load drift 2000 hours 75°C, ratings (%)	6	1		1	0.1	0.03	1
Noise voltage coefficient 10 kΩ resistor	10 μV/V	0.1 μV/V		0.03 μV/V	0.03 μV/V	White thermal noise	
Voltage coefficient	200 ppm/V	3 ppm/V		2 ppm/V	0.05 ppm/V	Not measurable	
Frequency response →1 kΩ →100 kΩ	→100 MHz →5 MHz	→250 MHz →50 MHz	→50 k Hz →50 k Hz	→250 MHz →50 MHz	→250 MHz →50 MHz		
Temperature coefficient	750 ppm/°C	250 ppm/°C	20 ppm/°C	100 ppm/°C	50 ppm/°C	5 ppm/°C	100 ppm/°C
Tolerances available (%)	5 → 20	1 → 10	0.01 →	0.5 → 5	1 → 5	0.005 →	0.5–5
Failure rate (FITS)	0.2	0.1	0.3	0.1	0.1	0.05	0.2

The inductance is mainly due to the lead, although wound capacitors also have an inductive component. Here a sandwich of foil electrodes and plastic dielectric film are rolled into a compact cylinder (see Figure 5.12). Some capacitors, particularly disc ceramics, are sold as low inductance types and are ideal for bypassing integrated circuit logic packages to eliminate noise on supply lines.

Because capacitors have an inductance component, there is a self-resonant frequency (see Figure 5.8) above which the capacitor has an inductive impedance. Lead lengths, components and track on a card must be kept short. For example, if a 1000 pF capacitor, frequently used in radio frequency circuits for earthing, has 100 mm of track or lead associated with it, the self-resonant frequency is under 20 MHz.

There are four principal dielectric materials used in capacitors. They are:

1. mica, which is gradually being replaced
2. plastic, the most common being polyester and polystyrene
3. ceramic
4. oxides, such as aluminum oxide and tantalum dioxide.

For the latter, a very thin, high dielectric oxide film is formed electrolytically, giving a large capacitor value in a small volume. Being made by an electrolytic process, the capacitors are usually DC polarity sensitive and must be inserted into the circuit with the correct polarity. Materials used are aluminum and tantalum. Non-polarized tantalum types are also available.

Important for capacitors are their loss and hysteresis characteristics, which depend upon the dielectric material used. Some are made for specific applications such as:

- Radio frequency bypass capacitors where the capacitor value is the guaranteed minimum value of the capacitor.
- Storage capacitors for sample and hold circuits (polystryrene, polypropylene or Teflon dielectric being the usual materials).

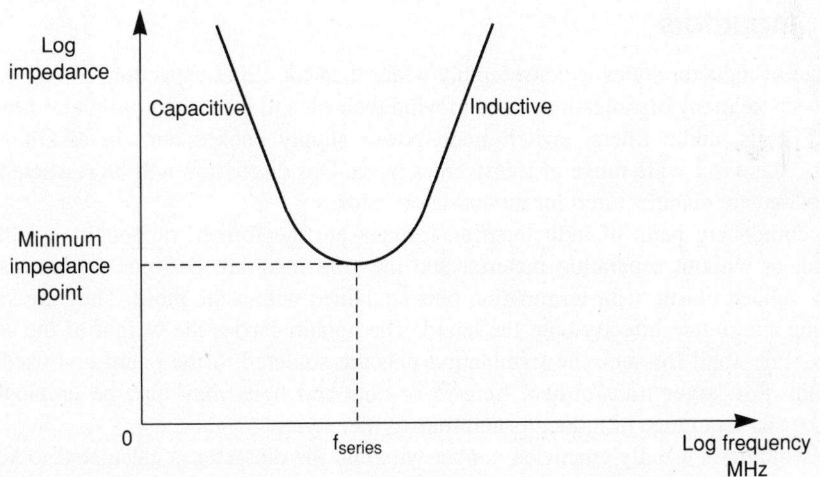

Figure 5.8 Impedance/frequency characteristic of a capacitor

Table 5.4 Characteristics of a 64 μF aluminum electrolytic capacitor at 25°C when new and after two years

Frequency	Impedance in ohms		
	Ideal	**New device**	**After 2 years**
50 Hz	50	50	50
1 kHz	2.5	2.6	3
100 kHz	0.025	1	1.6

- Ceramic capacitor types with positive and negative temperature coefficients to allow circuits such as tuned circuits in LC oscillators, to be balanced to have a zero temperature coeffficient. (The inductance normally has a positive temperature coefficient.)

Like resistors, capacitors have voltage ratings (DC and AC), temperature coefficients, stability and ageing characteristics. Electrolytic capacitors have the poorest performance of all types as illustrated in Table 5.4.

As the table shows, this capacitor is not recommended for use above audio frequencies. While tantalum capacitors can be used up to frequencies approaching 10 MHz, all electrolytic bypass capacitors should have a shunt ceramic or plastic capacitor in parallel with them to provide adequate high frequency bypassing.

Small variable trimming capacitors are required for some circuits; for example, the exact presetting of a crystal oscillator frequency. Like the variable resistors they are adjusted by screwdriver. There are two main types. The first has a set of meshing plates (180° turn) with either air or a plastic dielectric, and the second, two flat or concentric plates that are brought closer together by a screw. Often the dielectric is a mica or plastic sheet (see Figure 5.12).

Table 5.5 shows typical characteristics of a number of capacitor types.

5.2.3 Inductors

The range of inductor styles is considerably wider than for either capacitors or resistors. This is because many organizations need to wind their own to meet their particular needs, be it RF coils, audio filters, switch-mode power supply chokes, etc. In addition to inductors, there is a wide range of transformer types. Our discussion will be restricted to inductors that are manufactured for mounting on cards.

The component parts of inductors/transformers are the former, or bobbin, winding wire with or without separating material and the core material. Bobbins are normally made of molded plastic with termination pins contained within the mold. They are self-supporting and mount directly onto the board. The bobbin carries the weight of the wire and core. For small formers, the termination pins are soldered to the board and used as the mount. For larger transformers, screws or nuts and bolts may also be employed. Figure 5.9 shows a range of inductors and transformer styles.

The winding is usually enameled copper wire and the diameter is calculated to keep the temperature rise under full load to an acceptable level, as well as provide the correct quality factor (Q) and turns required for the given inductance value. Naturally, the wire

Table 5.5 Properties of several capacitor types

Parameter	Paper	Mica	Ceramic	Polystyrene	Polycarbonate	Teflon	Aluminum foil electrolytic	Tantalum electrolytic
Typical capacitance range (μF)	0.001–1	1 pF–0.1	5 pF–2		0.01–5	0.001–4	0.5–10,000	1–5,000
Typical tolerance (%)	± 20	± 5	± 10	± 10	± 10	± 10	0 → + 150	– 15 → + 75
Typical dissipation factor at frequency	0.4 1 kHz	< 0.1 1 kHz	0.01–2.5 1 kHz		1.0 1 kHz	1.0 1 kHz	5 50 kHz	1 → 10 50 kHz
Maximum insulation resistance 25°C	500 MHz 5000 plus	10 GHz 20,000 plus	5 GHz 100,000 plus	1 GHz	100 MHz 50,000 plus	1 MHz 100,000 plus	0.1 μA leakage current	0.05 μA leakage current
85°C	50 plus	2,000 plus	2,000 plus		2,000 plus	10,000 plus	0.5 μA leakage current	0.2 μA leakage current
Temperature operating range		–55 → 150	–55 →125		– 55 → 125	–55 → 250	–80 → 150	–80 → 150
Temperature coefficient		0 to + 70 ppm/°C	+ and – ve available		± 2.5%	± 1%		–
Stability (ageing)	Average	Very good	Good		Good	Average	Poor	Average
Failure rate (FITS)	0.2	2	0.3	0.5	0.5	0.5	10.0	1.0

must fit into the bobbin volume. The winding can be jumbled or scrambled, layered, interleaved, orthocyclic, bifilar, or wave. Some of these are illustrated in Figure 5.10.

Bifilar winding means all windings are wound simultaneously. It provides the maximum possible coupling between windings. Orthocyclic winding allows maximum wire in a given cross-sectional area. Wave winding, although complex to perform, provides the least stray capacitance. Alternatives to provide low capacity are scrambled or distributed winding. Distributed winding splits up winding so that the inductance consists of several separated coils in series. Commonly used for radio frequency chokes, this method is illustrated in Figure 5.11a. For VHF chokes, a common practice

Figure 5.9 Selection of inductance styles

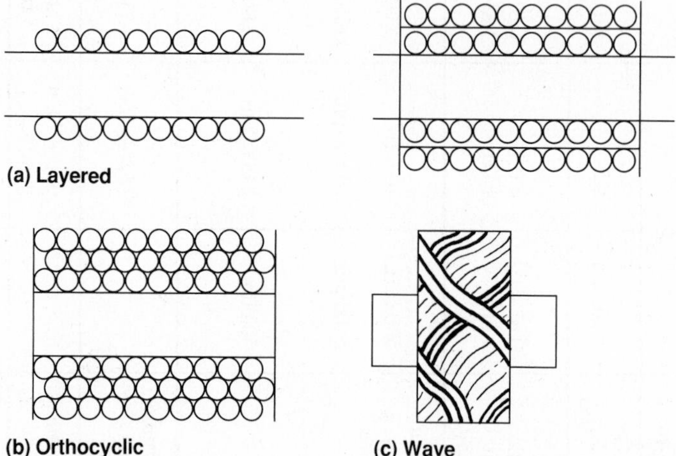

Figure 5.10 Inductor/transformer winding styles: (a) layered, (b) orthocyclic, and (c) wave

is to use a length of wire one-quarter wavelength long and to layer or jumble wind the wire onto a resistor in the range 3.3 k to 10 kΩ.

Core material can be laminated steel (audio frequencies only), powdered iron, a ferrite or occasionally a brass slug. When inserted into a coil, a brass slug reduces the inductance value whereas the others, being ferromagnetic, increase the inductance. Air core coils are frequently used at very high frequencies and/or power levels.

The shape of the core is also variable. Toroids are the most efficient, but are difficult to wind. Because some ferrites conduct electricity, molded ferrite toroids are often encased in plastic so rough ferrite edges will not damage the enamel of the winding wire, shorting turns. Steel strips can also be wound to form a toroid, but two C shaped cores (either steel or ferrite) clamped or glued together perform almost as well and allow a bobbin to be used. For radio frequency work, iron dust (up to typically 30 MHz) or ferrite slugs are screwed in or out of the coil to vary the inductance. The coils often have an aluminum shield placed around them to reduce the interaction with other items in close proximity (see Figure 5.11b). At audio frequencies special ferrite cup or core moldings are employed to surround the central bobbin. Holes in the ferrite allow ½ and ¼ turns to be used. If an air gap is used, a slug can be added for fine tuning (see Figure 5.11c). For surface mount work, special formers are appearing. Many are similar to the leaded ones, but a new type using molded ferrite is shown in Figure 5.11d.

To provide mechanical protection or reduce high voltage corona effects, inductors and transformers may be potted. Two problems arise. First, the potting material, no matter

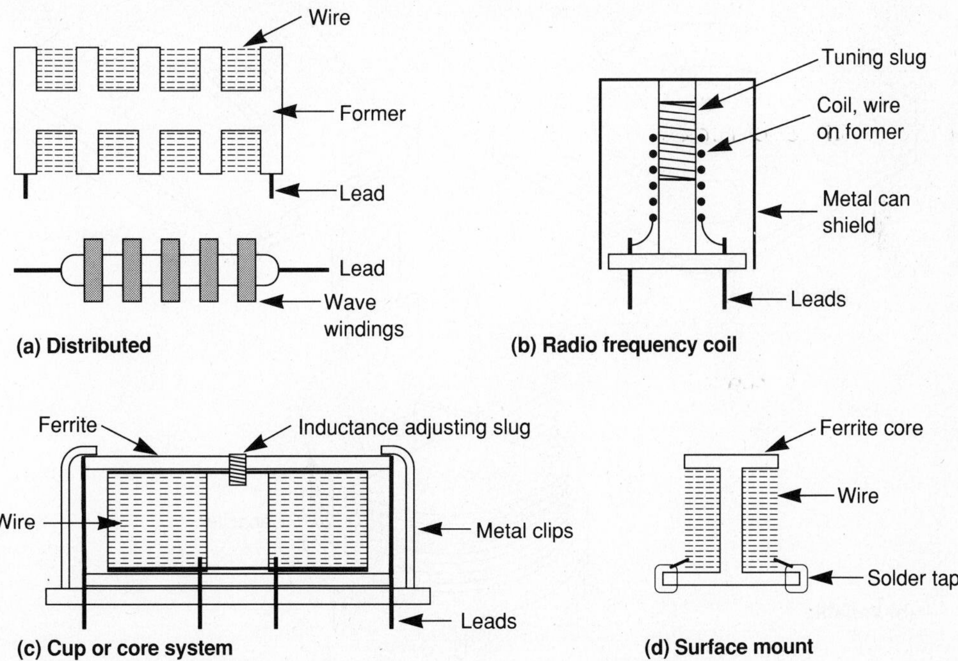

Figure 5.11 A selection of inductors: (a) distributed, (b) shielded radio frequency coil, (c) cup or core, and (d) surface mount ferrite inductor

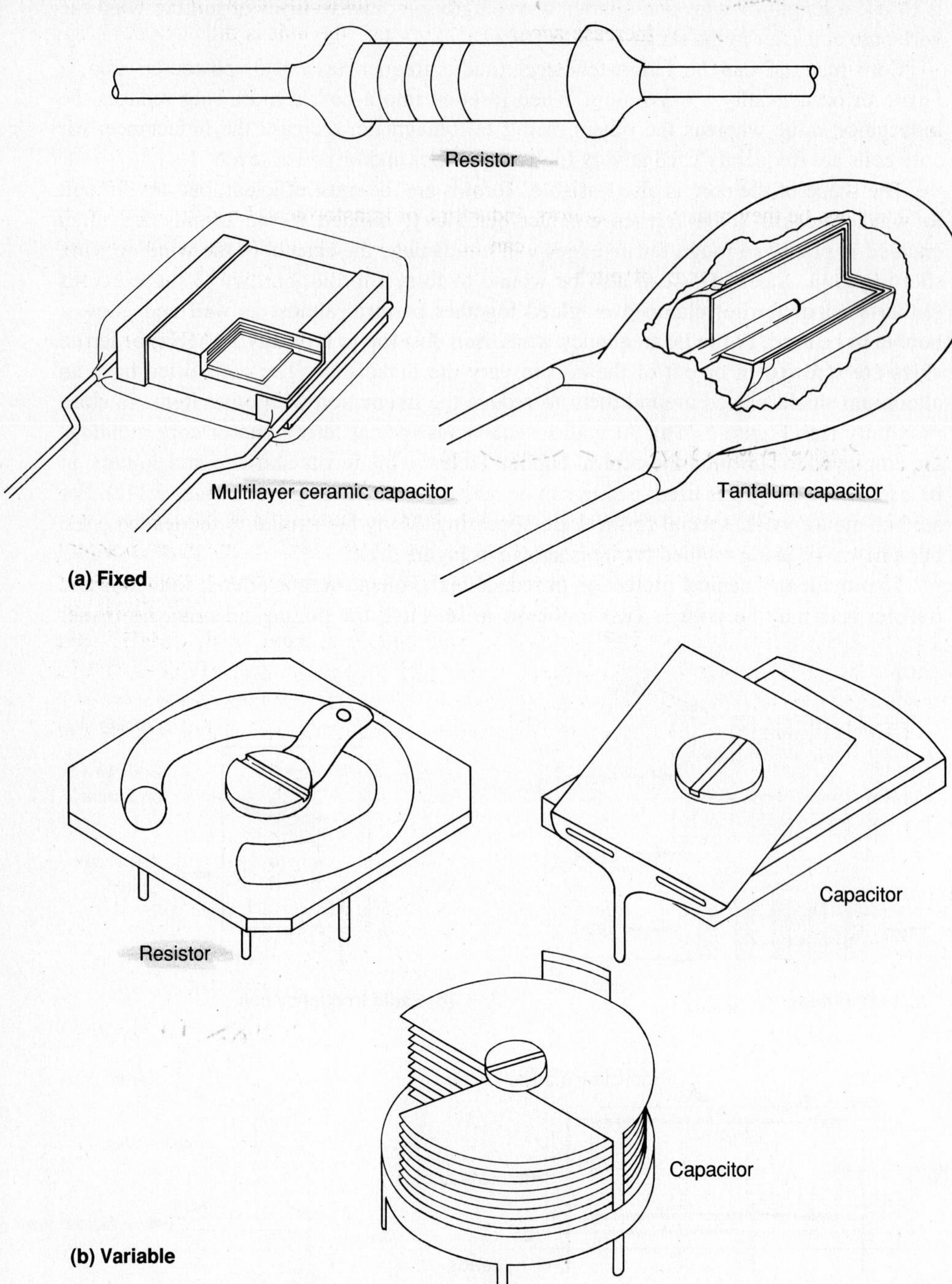

Figure 5.12 Leaded component devices: (a) fixed value components, and (b) variable value components

whether it is a phenolic, urethane or epoxy type, has a dielectric constant greater than unity, so all stray capacities increase accordingly. Second, the unit is difficult and often impossible to repair once it has been potted.

5.2.4 Packaging

At present there is no uniform standard that specifies the dimensions of passive components, be they resistors, capacitors, inductors, or transformers. Leaded components present little difficulty because the leads can be formed to provide stress relief and can accommodate a wide range of pitches on difficult layout grids. Figure 5.12 shows the construction of leaded resistors and capacitors.

For surface mount devices, the rectangular chip form is the norm. Figure 5.13 shows typical construction methods.

5.3 Semiconductor components

Like passive components, the semiconductor field is constantly evolving as new components and processes mature and replace the older ones. For example, germanium devices have been replaced by silicon; discrete devices by integrated circuits; diode transistor logic by transistor; transistor logic in turn is being replaced by CMOS; fusable ROMs by EPROMs and now EEPROMs; and 4 bit microprocessors by 8, 16 and 32 bit chips. Change is also occurring with the package styles used to contain these devices. The longstanding dual in line (DIL) package is now succumbing to surface mount techniques and, in the future, these may even be replaced by tape automated bonding (TAB) and chip-and-wire bond methods.

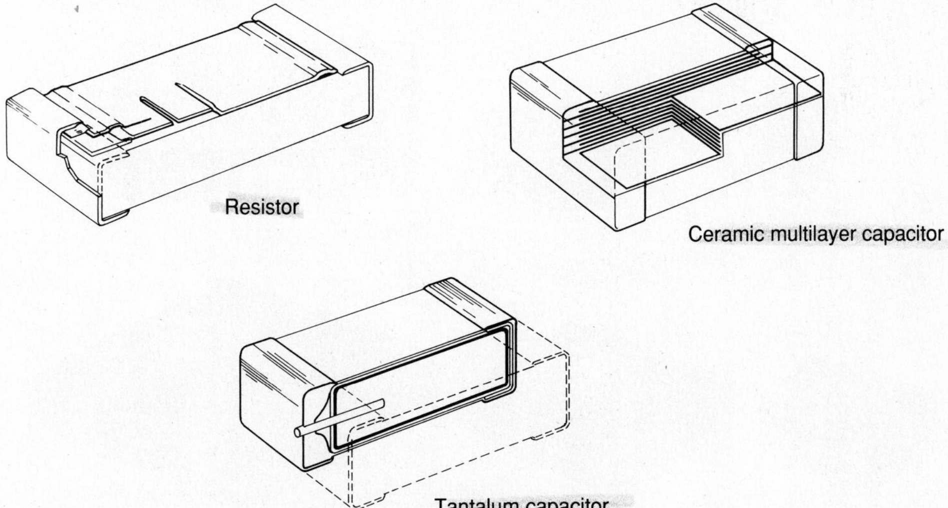

Figure 5.13 Surface mount device construction and shapes: (a) resistor, (b) ceramic multilayer capacitor, and (c) tantalum capacitor

For almost two decades the approach to assembly, using leaded semiconductor components, has not changed. The standard DIL packages with 100 mil spacing has worked well. Unfortunately, the packages have several limitations. First, the package is considerably larger than the semiconductor chip encapsulated. Thus, the large footprint is wasteful of card area. Second, leads towards the corner of the package have much higher series inductance and stray capacitance than those nearer the center. Figure 5.14 shows an assortment of semiconductor packages and highlights some of these problems.

Third, the 100 mil lead spacing placed a severe limit on the number of pins that could be accommodated in practical terms, with 64 being about the limit. The growth of VLSI and the pressure for miniature systems has forced new package types to appear and some older surface mount types to become popular. At present, the industry is in a state of turmoil, for as yet a dominant package type to replace the DIL has not yet appeared. For low count pin numbers, the small outline (SO) package with gull leads for surface mounting is popular (see Figure 5.15a). It is available in 3, 4, 8, 14, 16 and 28 pin types. Several versions are available. Pins are on a 50 mil grid.

Figure 5.14 Assortment of semiconductor packages. Note the difference in lead lengths between package types

For higher pin counts, the quad package with J leads has also become an accepted package in the industry. Quad packages have the advantage that they are only the central core of a dual in line package, thereby giving a smaller footprint and more constant impedance per lead. This is illustrated in Figure 5.16.

When extremely large pin counts are needed, the pin grid array is used. It is not really suitable for surface mount boards unless a surface-mount socket is employed.

Comparing these three basic package styles, Table 5.6 shows the number of pins or leads available.

Because the gull leads on a small outline package protrude well beyond the package, they can be easily damaged. But the fact that they do protrude means that they can be probe tested and inspected easily. The J lead presents problems for inspection and achieving correct orientation. It is difficult to see under the package to check for any solvent residue. Further, to achieve good soldering, the planarity of the board must be held to within 2 mil (50 microns). Pin grid arrays are seen only as a temporary solution because they are extremely expensive and pose difficulties in the layout of the card on

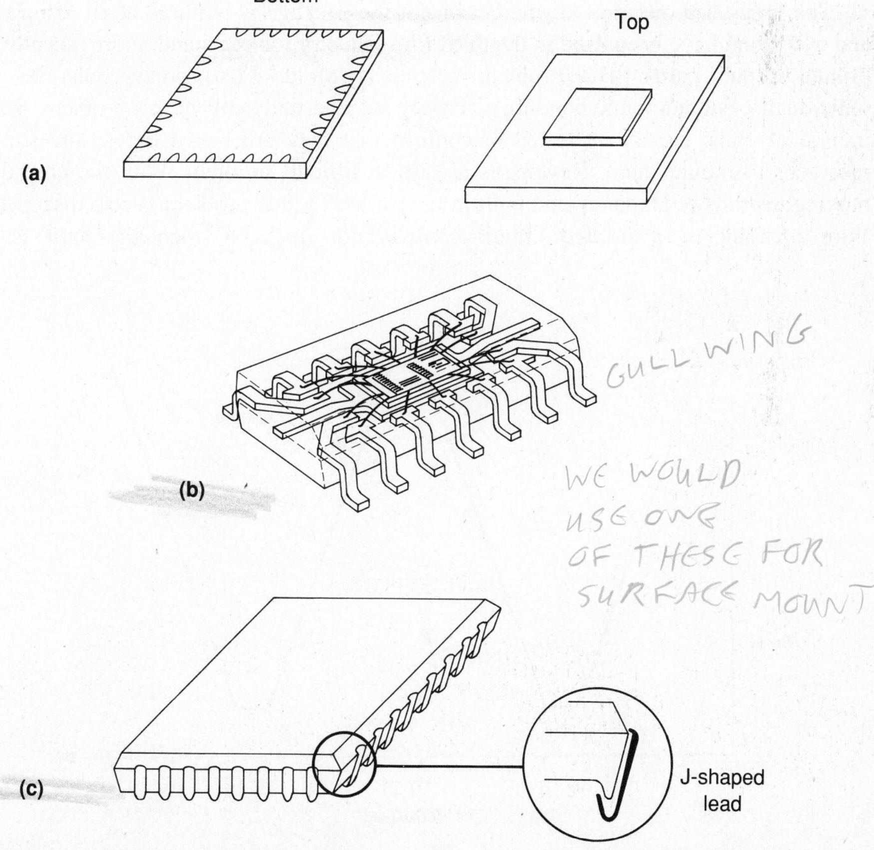

Figure 5.15 Surface mount package types: (a) JEDEC leadless chip carrier, (b) small outline gull wing, and (c) Texas Instruments J lead

Table 5.6 Pin counts available for various package styles

Style	Number of pins/leads	Surface mount type
DIL	4 to 64	No
Small outline	3 to 28	Yes
J lead	24 to 88	Yes
Pin grid array	64 to 300 plus	No

which they are to be mounted. Other solutions abound and, before the design of any card, package configurations and footprints need to be confirmed. It is customary in the United States to maintain a constant lead separation for a range of packages, be it 50, 40, 25 or 20 mil spacing. On the other hand, the Japanese vary their pitch. A recent range of Canadian packages from Northern Telecom does the same. The C Quad package has the same overall dimensions, with the pitch decreasing in steps with lead count from 50 down to 30, then to 25 mil.

The long-term question is whether or not the package is required at all. Chip mount and wire bond have been used in the thick film industry for years and, more recently, with printed circuit cards. The problem here is twofold. First, bonds must be made individually, but automatic bonding machines have partially solved this problem. Second, individual chips cannot be tested to confirm that they are sound before investing the resources to mount them. Rework is always a difficult problem. With the chip design moving towards redundancy and built in testing (BIT), this problem is also disappearing. Wire bonding to a printed circuit card is not difficult when the card is new.

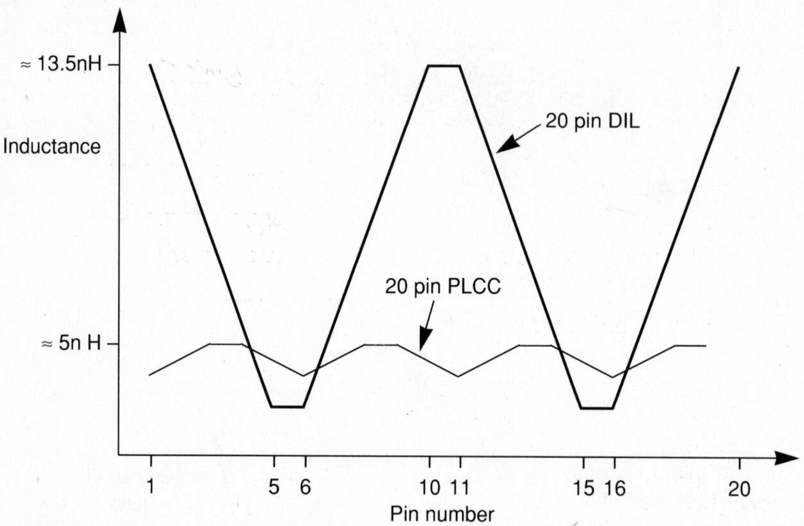

Figure 5.16 Variation in lead inductance for two package types (Texas Instruments 1988)

With the passage of time and oxidation, bonds become a problem. The solution is to apply a 0.2 micron gold flash over the copper or, preferably, a nickel layer first before the gold flash. Standard hard gold can be used. When the bond is made, the gold vaporizes and the bond is made to the clean material underneath.

An alternative is the tape automatic bonding system (TAB), in which chips are gang bonded to a copper lead frame, mounted on a polyester or polyimide movie-film-like support. Initially this system had problems because each chip required a different lead frame and chips needed to be bumped before being joined to the frame. Today, standard lead frame sizes are being made and the lead frame itself is bumped. The chip can be encapsulated in plastic or left exposed and protected by its own glaze. Normally, when mounted, the rear of the chip is facing upwards. Figure 4.4 illustrates the chip-and-wire bond assembly method and Figures 1.11 and 5.17 show TAB.

5.4 Wires

An important component in electronic assemblies is the wire used to manufacture and interconnect components. Wire may be classified under several headings, including:

- interconnection or 'hook up' wire
- enamel coated (magnet) wire
- Litz wire
- tinned copper wire
- coaxial and shielded cables/wires

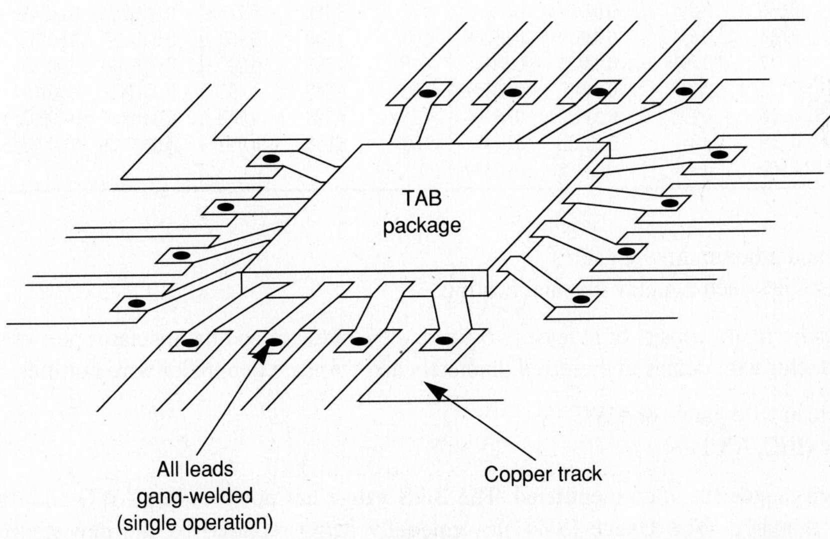

Figure 5.17 TAB, an alternative mounting arrangement to bonding

Table 5.7 Metric bare copper wire (from Australian Standard AS 1194.1)

Conductor diameter (mm)			Conductor resistance Ω/m at 20°C		Conductor diameter (mm)			Conductor resistance Ω/m at 20°C	
Nominal	Max.	Min.	Max.	Min.	Nominal	Max.	Min.	Max.	Min.
0.020	0.022	0.018	68.26	44.03	0.950	0.960	0.940	0.02503	0.02312
0.025	0.027	0.023	41.81	29.23	1.000	1.010	0.990	0.02256	0.02089
0.032	0.034	0.030	24.57	18.44	1.06	1.071	1.049	0.02009	0.01857
0.040	0.042	0.038	15.32	12.08	1.12	1.131	1.109	0.01798	0.01666
0.050	0.052	0.048	9.599	7.881	1.18	1.192	1.168	0.01621	0.01499
0.063	0.065	0.061	5.944	5.044	1.25	1.263	1.237	0.01445	0.01335
0.071	0.074	0.068	4.783	3.892	1.32	1.333	1.307	0.01295	0.01199
0.080	0.083	0.077	3.730	3.095	1.40	1.414	1.386	0.01151	0.01065
0.090	0.093	0.087	2.922	2.464	1.50	1.515	1.485	0.01002	0.009285
0.100	0.103	0.097	2.351	2.009	1.60	1.616	1.584	0.008814	0.008160
0.112	0.115	0.109	1.862	1.611	1.70	1.717	1.683	0.007808	0.007228
0.125	0.128	0.122	1.486	1.301	1.80	1.818	1.782	0.006964	0.006447
0.140	0.143	0.137	1.178	1.042	1.90	1.919	1.881	0.006250	0.005787
0.160	0.163	0.157	0.8973	0.8021	2.00	2.020	1.980	0.005641	0.005222
0.180	0.183	0.177	0.7060	0.6363	2.12	2.141	2.099	0.005019	0.004649
0.200	0.203	0.197	0.5698	0.5171	2.24	2.262	2.218	0.004495	0.004165
0.224	0.227	0.221	0.4528	0.4135	2.36	2.384	2.336	0.004052	0.003749
0.250	0.254	0.246	0.3654	0.3303	2.50	2.525	2.475	0.003610	0.003342
0.280	0.284	0.276	0.2903	0.2642	2.65	2.677	2.623	0.003214	0.002973
0.315	0.319	0.311	0.2286	0.2094	2.80	2.828	2.772	0.002878	0.002664
0.355	0.359	0.351	0.1795	0.1654	3.00	3.030	2.970	0.002507	0.002321
0.400	0.405	0.395	0.1417	0.1299	3.15	3.182	3.118	0.002274	0.002104
0.450	0.455	0.445	0.1116	0.1029	3.35	3.384	3.316	0.002011	0.001861
0.500	0.505	0.495	0.09026	0.08356	3.55	3.586	3.514	0.001791	0.001657
0.560	0.566	0.554	0.07206	0.06652	3.75	3.788	3.712	0.001605	0.001485
0.630	0.636	0.624	0.05679	0.05268	4.00	4.040	3.960	0.001410	0.001305
0.710	0.717	0.703	0.04475	0.04145	4.25	4.293	4.207	0.001249	0.001156
0.750	0.758	0.742	0.04017	0.03709	4.50	4.545	4.455	0.001114	0.001031
0.800	0.808	0.792	0.03525	0.03264	4.75	4.798	4.702	0.0010003	0.0009257
0.850	0.859	0.841	0.03126	0.02888	5.00	5.050	4.950	0.0009026	0.0008356
0.900	0.909	0.891	0.02785	0.02579					

- ribbon and other multiwire cables
- special cables such as delay line and heating.

The wires normally consist of at least two parts, the conductor and the insulator portions. The conductor wire comes in preferred diameters and there are two major wire families:

1. American wire gauge or AWG
2. metric (IEC, AS 1194.1).

Two other gauges are often mentioned. The B&S gauge has now become AWG and the Imperial Standard Wire Gauge (SWG) is gradually being replaced by the new metric sizes.

Tables 5.7 and 5.8 show a range of copper wire gauges.

The properties for several different wire types will now be considered in detail.

Table 5.8 Comparison of metric and gauge wire diameters. Diameters differing by not more than 1% are considered identical (Magnet Wire 1973)

Australian Standard * = 1st Preference		Gauge B&S (AWG)	SWG	Non-standard diameters	
mm	inch			mm	inch
* 0.025	0.001 0	50	50		
0.028	0.001 1	49			
			49	0.030	0.001 2
* 0.032	0.001 3	48			
0.036	0.001 4	47			
* 0.040	0.001 6	46			
			48	0.041	0.001 6
0.045	0.001 8	45			
* 0.050	0.002 0	44	47		
0.056	0.002 2	43			
			46	0.061	0.002 4
* 0.063	0.002 5	42			
0.067	0.002 6				
* 0.071	0.002 8	41	45		
0.075	0.003 0				
* 0.080	0.003 1	40			
			44	0.081	0.003 2
0.085	0.003 4				
* 0.090	0.003 5	39			
			43	0.091	0.003 6
0.095	0.003 7				
* 0.100	0.003 9	38			
			42	0.102	0.004 0
0.106	0.004 2				
* 0.112	0.004 4	37	41		
0.118	0.004 7				
			40	0.122	0.004 8
* 0.125	0.004 9				
		36		0.127	0.005 0
0.132	0.005 2		39		
* 0.140	0.005 5				
		35		0.142	0.005 6
0.150	0.005 9				
			38	0.152	0.006 0
* 0.160	0.006 3	34			
0.170	0.006 7		37	0.173	0.006 8
* 0.180	0.007 1	33			
0.190	0.007 5				
			36	0.193	0.007 6
* 0.200	0.007 9				
		32		0.203	0.008 0
0.212	0.008 4		35		
* 0.224	0.008 8	31			
0.236	0.009 3		34		
* 0.250	0.009 8				
		30	33	0.254	0.010 0
0.265	0.010 4				
			32	0.274	0.010 8
* 0.280	0.011 0				
		29		0.287	0.011 3
			31	0.295	0.011 6
0.300	0.011 8				
* 0.315	0.012 4		30		
		28		0.320	0.012 6
0.335	0.013 2				
			29	0.345	0.013 6

Australian Standard * = 1st Preference		Gauge B&S (AWG)	SWG	Non-standard diameters	
mm	inch			mm	inch
* 0.355	0.014 0				
		27		0.360	0.014 2
0.375	0.014 8		28		
* 0.400	0.015 7	26			
			27	0.417	0.016 4
0.425	0.016 7				
* 0.450	0.017 7	25	26	0.455	0.017 9
0.475	0.018 7				
* 0.500	0.019 7				
		24	25	0.511	0.020 0
0.530	0.020 9				
* 0.560	0.022 0		24		
		23		0.574	0.022 6
0.600	0.023 6				
			23	0.610	0.024 0
* 0.630	0.024 8				
		22		0.643	0.025 3
0.670	0.026 4				
* 0.710	0.028 0		22		
		21		0.724	0.028 5
* 0.750	0.029 5				
* 0.800	0.031 5				
		20	21	0.813	0.032 0
* 0.850	0.033 5				
* 0.900	0.035 4				
		19	20	0.920	0.035 9
* 0.950	0.037 4				
* 1.000	0.039 4				
		18	19	1.024	0.040 3
* 1.060	0.041 7				
* 1.120	0.044 1				
		17		1.151	0.045 3
* 1.180	0.046 5				
			18	1.219	0.048 0
* 1.250	0.049 2				
		16		1.290	0.050 8
* 1.320	0.055 1				
			17	1.422	0.056 0
		15		1.450	0.057 1
* 1.500	0.059 1				
* 1.600	0.063 0				
		14	16	1.628	0.064 1
* 1.700	0.066 9				
* 1.800	0.070 9				
		13	15	1.828	0.072 0
* 1.900	0.074 8				
* 2.000	0.078 7				
		12	14	2.052	0.080 8
* 2.120	0.083 5				
* 2.240	0.088 2				
		11		2.304	0.090 7
			13	2.337	0.092 0
* 2.360	0.092 9				
* 2.500	0.098 4				
		10		2.588	0.101 9
* 2.650	0.104 3		12		

Source: Magnet Wire Pty Ltd. Reproduced with permission

5.4.1 Interconnect or 'hook up' wire

This is the general purpose wire used for interconnecting cards and components. The conductor can be a single strand or multistrand. The latter must be employed where there is vibration or mechanical movement. The wire for the multistrand is made up of concentric layers or strands, as illustrated in Figure 5.18. The greater the number of strands for a given diameter, the more flexible and vibration resistant. It is also more expensive. Typical hook up wire is 7 and 14 strands. The wire is usually drawn, annealed copper which may or may not be tin plated.

Copper is hard drawn through a range of decreasing diameter dies, each drawing reducing the diameter of the wire until it reaches the required size. Other wire materials used include aluminum and stainless steel (strength at small diameters).

The wire is coated with an insulator which is normally a plastic, the most common being polyvinyl chloride (PVC), Teflon (TFE) and Kynar (PVF). The PVC is for general purpose use, the more expensive Teflon for high temperature operation and the Kynar, with its abrasive qualities and cut-through resistance, for wire-wrap connections. Table 5.9 summarizes the properties for these three types of insulation.

Special insulated wires are also available, but they are expensive. An example is glass impregnated nylon, which has a breakdown voltage of several tens of kilovolts.

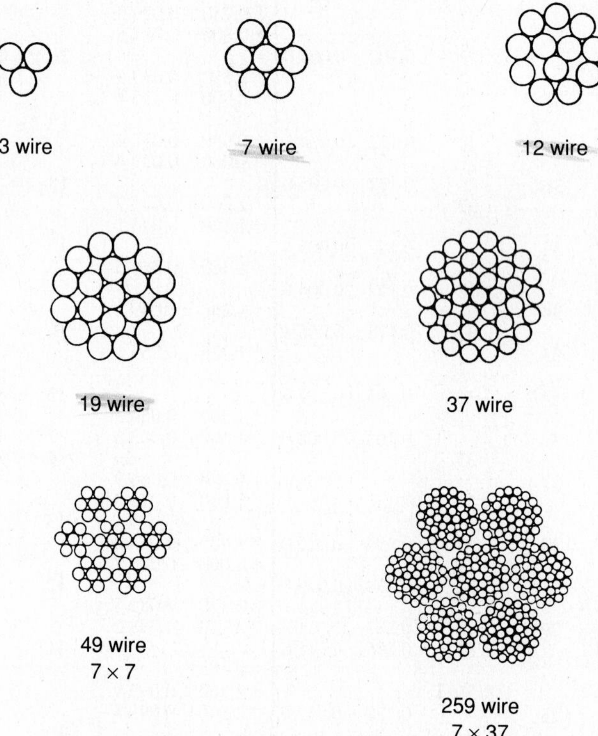

Figure 5.18 Selected sections showing the assembly of multistrand wire

Table 5.9 Common wire insulators used in the electronics industry (PVC is for general purpose, TFE high temperature operation and Kynar for wire wrap)

Common name	Full name	Fluid solvent resistance	Flammability	Temperature range °C	Abrasion	Cut through resistance	Dielectric strength V/mil	Dielectric constant 1 kHz	Loss factor 1 kHz
PVC	Polyvinyl chloride	Good	Slow burning	$-55 \rightarrow 105$	Poor	Poor	400	5–7	0.02
Teflon (TFE)	Polytetrafluoro ethylene	Ex	Non-flammable	$-80 \rightarrow 260$	Fair	Fair	480	2.1	0.0003
Kynar	Polyvinylidine fluoride	Good	Self-extinquishing	$-65 \rightarrow 130$	Good	Good	160	7.7	0.02

Table 5.10 Summary of commonly used enamels

Name	Properties	Use
Oleo-resinous	Original type: low abrasion resistance; attacked by some solvents (e.g., acetone, trichlorethelyne, carbon tetrachloride)	
Polyvinyl formal (in the polyvinyl acetal family)	Good all round properties: abrasion resistance good; flexibility; good adhesion to copper; resistant to solvents, transformer oils and freon; temp. → 120°C	Motors Transformers Generators
Polyurethane (solderite) pink color	Good all round as above. Can solder to wire direct; temp. → 120°C; with nylon coating (overcoat) → 130°C	RF coils Small motors Solenoids/relays
Polyester	Good all round as above except attacked by some solvents (e.g., acetone and trichlorethylene; temp. higher → 155°C	Transformers Relays/solenoids High performance Motors
Polyesterimide	As above but temp. → 180°C	
Polyamide-imide	Good properties: highest temp. → 220°C. Must remove all water (i.e., use in a hermetically sealed environment)	High temp. relays/motors

Table 5.11 Summary of enamel wire overcoats

Material	Properties
Nylon	Improves overall performance; resistance to solvents; abrasion: use increased winding speeds
Textile covering	Rayon, cotton, nylon, silk; absorbs wax
Heat bond wire	Thermoplastic or thermosetting coating; heat by passing current through wire
Solvent bonding	Thermoplastic material coated wires can be set with solvents such as isopropanol alcohol or methanol alcohol Acetate rayon coating with acetone

When selecting the correct interconnection wire for assembly, the following points need to be considered:

- resistance per unit length: the voltage drop along a wire for the current flowing in it
- self-heating of the wire and the thermal characteristics of the insulation
- fusion current: the maximum current the wire can pass
- insulation voltage
- environmental conditions: temperature range, any chemicals or fluids present, vibration, etc.

5.4.2 Enamel wire (or magnet wire)

This normally consists of a solid copper conductor with a lacquered enamel coating. Silk, cotton, nylon, or rayon fiber may be placed over the enamel to give it additional protection. Like the 'hook up' wire, the copper is hard drawn to size and then annealed. The enamels are normally plastic and may consist of layers of different materials to improve the overall properties for a particular application. The enamel wires are generally used for coils (for example, inductances, relays and solenoids, loud speaker coils and motors) covering a wide frequency, voltage and operating temperature range. Thus properties such as dielectric constant, smoothness of surface, maximum operating temperature, breakdown voltage and flexibility need to be considered.

The wire may be coated with thermoplastic or thermosetting materials so that, after a close spaced coil is wound, it may be heated to remain permanently in the wound position. The thermoplastic material can also be bonded with solvents such as isopropyl alcohol or methanol. The same can be done with wires that have an overcoat of nylon. If the overcoat is acetate rayon, then the solvent is acetone. Table 5.10 lists some of the common enamels used, giving their properties and typical applications. The polyurethane enamel has a characteristic pink tinge and is commonly used for radio frequency coils because the enamel can be removed by heat, allowing the wire to be soldered directly without having to mechanically strip off the enamel. Table 5.11 gives the types and properties of the common overcoats that are used to give added protection to the enamel wire.

5.4.3 Litz wire

As the frequency of operation of a circuit is increased, the current flowing in a wire concentrates at the surface. Thus the effective resistance of the conductor increases with frequency. This is called the 'skin effect'. By using several individually insulated wire strands in parallel (5–20), the effective surface area of the conductor can be increased. The strands must be threaded so that each wire has the same length at the outside of the bundle for any given wire length. This is to ensure that the current flowing will divide equally between the strands.

With radio frequency coils it is often important to have a high quality factor (Q) and the use of Litz wire over the frequency range 0.3–3 MHz offers significant improvement. Below 0.3 MHz the skin effect is of no real consequence and above 3 MHz the performance is not appreciably better than a single strand of enamel copper wire of the same overall diameter.

Commercial Litz wire is frequently made by simply twisting several strands of enameled copper wire together. While this is inferior to true Litz wire, it is a fair compromise between cost and performance. Because Litz wire is used for radio frequency coils, the strands are often given a textile overcoat and a wax so that, when wound, the wire will not slip or move on the former.

5.4.4 Tinned copper wire

As the name implies, tinned copper wire is a hard drawn, annealed copper that has received a coating of tin. It comes in various standard diameters (see Table 5.8) and has a wide range of applications including radio frequency coils and straps on printed circuit cards.

5.4.5 Coaxial and shielded cables/wires

Here the inner conductor(s) is surrounded by an outer one called the 'shielding'. The shielding may be made of conductors other than copper and can be:

- solid
- plaited braid
- parallel wires running the length of the cable
- wrapped foil.

By controlling the dimensions of the inner wire and outer shield cable and, correct selection of the dielectric material between the wires, cables may operate with a known

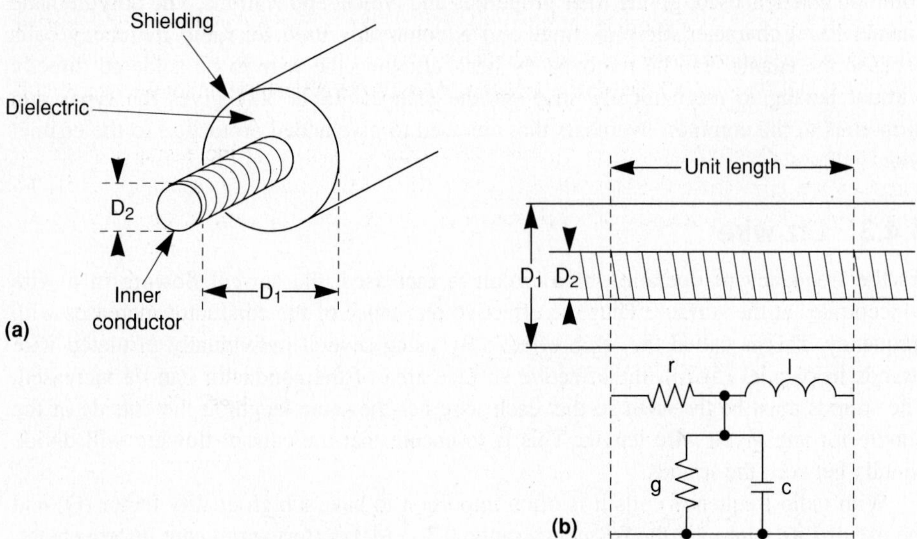

Figure 5.19 (a) Typical single inner conductor shielded cable, and (b) the model for a small section of such a cable

impedance over a wide frequency range (see Figure 5.19). The impedance is given by:

$$Z_0 = \sqrt{\frac{r + j\,\omega\,l}{g + j\,\omega\,c}} \qquad [5.2]$$

where r is the series resistance per unit length
l is the series inductance per unit length
g is the shunt conductance per unit length
c is the shunt capacitance per unit length.

For good-quality conductor and dielectric materials, the loss factors ‘r’ and ‘g’ can be ignored. Thus:

$$Z_0 \approx \sqrt{\frac{l}{c}} \qquad [5.3]$$

and is frequency independent. Used at radio frequencies, these cables are called coaxial cables. The impedance can also be expressed as a ratio of the inner and outer diameters (see Figure 5.19):

$$Z_0 = 138 \log_{10} D_1/D_2 \qquad [5.4]$$

The delay introduced by the shielded wire is given by the propagation constant:

$$\gamma = \sqrt{(r + j\,\omega\,l) \,.\, (g + j\,\omega\,c)}$$

$$\approx \omega \sqrt{l\,c} \qquad [5.5]$$

An important part of the loss is the leakage through the shielding (radiation). Thus a solid shield is best with the performance of braid and parallel wires falling off with frequency as the dimensions of the holes in the shielding become increasingly greater in terms of fraction of the wavelength of operation. The best dielectric is air and yet there needs to be a support for the inner conductor. Some coaxial cables use solid dielectric while others use a twisted helix of dielectric along the cable length.

Audio shielded wire is similar in construction to coaxial cable, but uses lower quality dielectric material and braid.

Delay or filter cable is a special case of a coaxial cable. By increasing l and c, the propagation constant and hence delay can be made very large (see Equation 5.5). The inner conductor is normally a continuous coil of enamel wire wound on a ferrite rod,

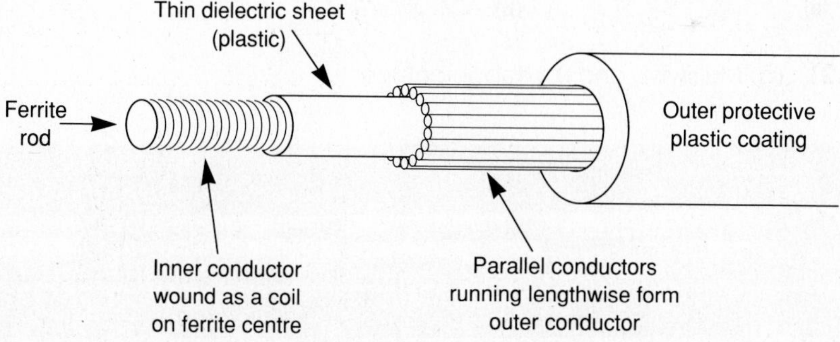

Figure 5.20 Construction of delay cable

resulting in a high l while the 'paper' thickness of the plastic dielectric ensures a large c per unit length. This is illustrated in Figure 5.20. The impedance Z_0 of these delay cables is high, being in the region of 600–2500 ohms.

5.4.6 Ribbon and multiwire cables

Frequently, several wires must run in parallel from one point to another. While a group of individual wires loomed together may be used, this method is not cost effective. Premanufactured cables, both circular and flat in cross section, are available. The circular consists of a bundle of multistrand wires with an outer plastic sheath. They may even be shielded. Ribbon cable is composed of multistrand wires side by side on spacing of the order of 60 mil (1.5 mm) and covered with a plastic molding. The insulating molding can be PVC, Teflon, or silicone, which is used to achieve greater flexibility. The ribbon can be purchased in various widths and is usually color coded to identify one side from the other. Figure 5.21 shows typical cross sections. Special insulation piercing connectors are used with flat ribbon for eliminate the need for strip insulation and soldering.

5.4.7 Looming of cables

Although ribbon cables and other techniques are gradually replacing looming in equipment, there are still many occasions when it is necessary. Looming is the lacing of cables together into appropriate bunches so that the product not only looks neat and tidy, but the hazard of catching loose wires in equipment drawers and cards is removed. It is an art, which originally was carried out using string soaked in wax for the looming material. Today it is common to use either PVC plastic flat, rod or tubing material. Figure 5.22 provides examples. The start and finish loops are usually double. Notice that at a loop, the loom material enters and leaves the knot from the bottom. The number of loops depends upon the size of the loom, wire diameter and the number of wire take-off points.

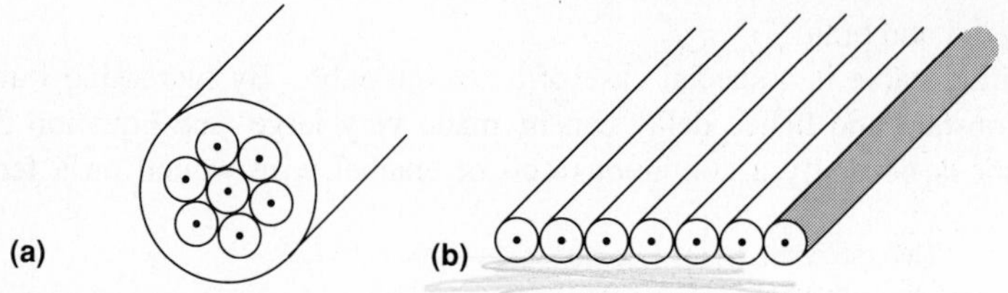

Figure 5.21 (a) Multiwire, and (b) ribbon cables

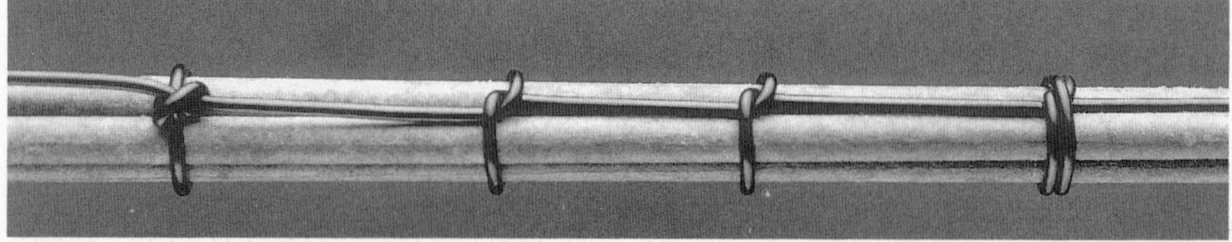

Figure 5.22 Example of looming

In a one-off or prototype situation, looming is usually undertaken *in situ* after the wiring has been completed. Where more product numbers are needed, a jig is normally made and the looming undertaken externally. As shown in Figure 5.23, the jig normally consists of marked timber with nails driven into it, the wires wrapped around the pins, loomed and then cut. To determine or trace a wire in a loom, several methods can be used. They include:

- using different colored wires
- using trace wires where the wire has a spiral trace of a second color on it

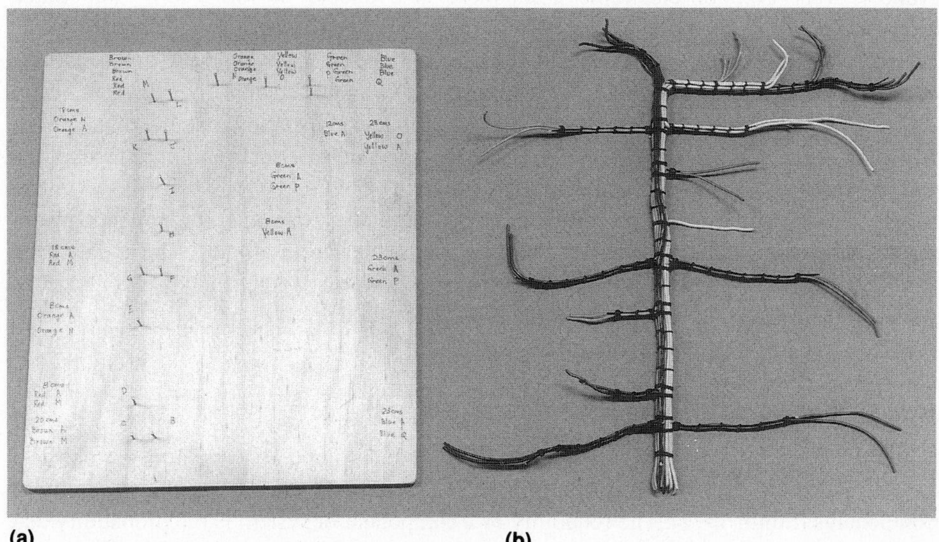

(a) **(b)**

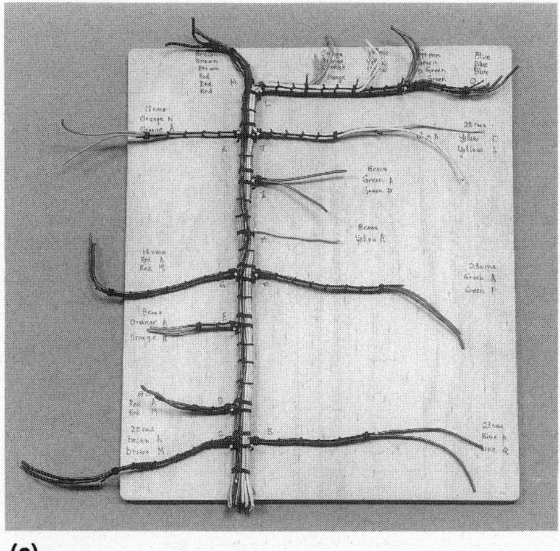

(c)

Figure 5.23 (a) Looming jig, (b) completed loom, and (c) loom before removal from jig

- identifying beads that clip on both ends of a wire. The beads can be color coded and/or have numbers on them.

Alternatives to conventional looming include taping, straps, 'zipper' tubing, and ties. In large volume equipment other construction methods can be used which eliminate wires altogether. These include mother-daughter board and flexible printed-board methods.

5.5 Other components

Almost every type of electronic component is available in some form for leaded mounting onto a printed circuit card. These include battery cases, loud speakers, ceramic filters, relays, switches, and connectors. Today more and more of these are becoming available in surface mounted forms. Some, like large connectors, have inherent difficulties so the normal leaded methods are still preferred by many manufacturers. What is certain is the gradual changeover, because the advantage of surface mount technology cannot be ignored.

Parallel with this change is the increasing complexity of some components. Wafer scale integration, multichip modules and hybrid modules are examples. The skills of the card designer and assembler will be taxed to the limit to ensure that such components can be mounted reliably and protected for the life of the product.

5.6 Component reliability

The quality and reliability of any product can only be as good as its component parts. In recent years, considerable effort has been made to gather information on the reliability of components (Smith 1972). The reliability of a component or system is the probability that it will perform its required function under given conditions for a specified operating time.

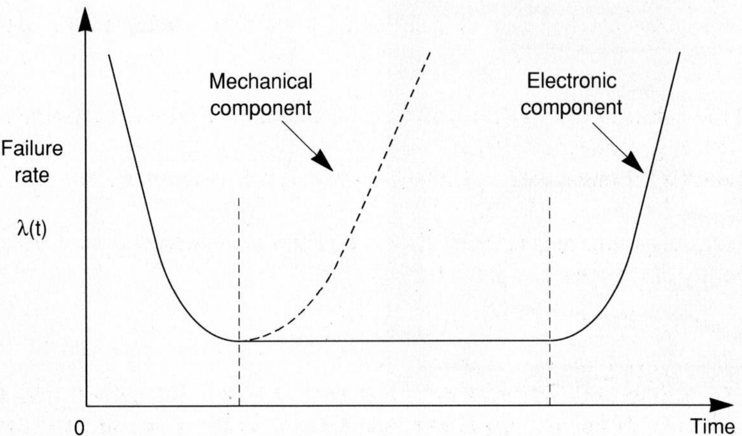

Figure 5.24 Failure rate of components with time

Thus the reliability R, is the probability of non-failure and has a value that will be between zero and one. When R approaches unity, it means that the system is very reliable (US Dept. of Defense MIL-HDBK-217E; Australian Standard 1211).

Conversely, the probability of failure or unreliability Q is given by:

$$Q = 1 - R \qquad [5.6]$$

While the reliability of a product depends upon many factors, including the design itself, manufacturing process, correct use of the product and adequate maintenance, the quality of the component is extremely important.

If a batch of components is subjected to normal operating conditions and the failure rate, defined as the number of failures in a given time, is plotted against time, it will result in what is often called the 'bath tub' curve, as shown in Figure 5.24.

With electronic components (heavy line) there are three distinct regions:

1. Infant mortality region or early failure period due to faulty components failing under normal stress. It is often a debugging period and the burn in process is used to take equipment through this region.
2. Constant failure rate region, where failures are random in nature and sudden. This is the normal service life for electronic equipment.
3. The wear out region where ageing of components, mechanical wear, drift in values cause failure.

For mechanical parts such as relays, plugs, sockets, and motors, the wear out is occurring throughout the entire life of the component so the dashed curve in Figure 5.24 applies.

The curve can be approximated by the Weibull function:

$$\lambda(t) = \beta.\lambda^{\beta}\, t^{(\beta-1)} \qquad [5.7]$$

where $\lambda(t)$ is the failure rate
t is time
β is a constant, being the slope of plots of cumulative failures with respect to time.

For: $\beta < 1$ the failure rate decreases with time—that is, the infant mortality region
$\beta = 1$ the failure rate is constant
$\beta > 1$ the failure rate increases with time—that is, the wear out region.

For the normal operational region of electronic equipment, the failure rate is a constant and the reliability of the system is given by:

$$R(t) = e^{-rt} \qquad [5.8]$$

where r is the failure rate. Its units are FITS (which stands for Failure unITS) and represent the number of failures in 10^9 component hours. In some literature the unit used is %/1000 hours or number of failures in 10^5 hours. It is this failure rate figure that is needed for each component so that the reliability of a product can be calculated.

If a system consists of m components of reliability R_{c1}, R_{c2} ... R_{cm} with respective failure rates of r_{c1}, r_{c2} ... r_{cm}, and if failure of one component means failure of the system (a series system), the reliability of the system R_s is given by:

$$R_s = R_{c1}.R_{c2}. \dots R_{cm} = \prod_{i=1}^{m} R_{ci} \qquad [5.9]$$

From Equation 5.4 we obtain:

$$R_s = e^{-t\left(\sum_{i=1}^{m} r_{ci}\right)} \qquad [5.10]$$

and the failure rate of the system is therefore:

$$r_s = \sum_{i=1}^{m} r_{ci} \qquad [5.11]$$

If, for example, a system had 10^5 identical components, each of a failure rate of 1000 FITS, and we require the equipment to operate eight hours a day, we can calculate:

1. the probability of the system being failure free for an eight-hour period.
2. the mean free time between failures.

From Equation 5.10 we have:

$$R_s = e^{-0.8} = 0.45$$

Thus, there is a 45% probability the eight hours would be failure free.

From Equation 5.11 we have:

$$r_s = 0.1/\text{hour}$$

Thus, the mean time between failures (MTBF) is 10 hours.

To estimate system reliability, it is important to know component failure rates. The same figures also allow the designer to select superior components for critical places to improve overall performance. Figures are normally supplied by large government organizations (defense, space), telecommunication companies and insurance organizations. Table 5.12 gives typical figures for failure rate of some components.

These figures are for normal conditions. Should the system be operated under harsh conditions, weighting or multiplying factors must be used. Examples are elevated temperatures (λ1), component power ratings (λ2), environment whether airconditioned, vehicular or missile (λ3). Thus the failure rate of a component taken from a table must be scaled by these factors, λ1 λ2 λ3, and these figures, derived from appropriate look-up tables. Table 5.13 gives some weighting factors.

Table 5.12 Typical failure rates for some components

Component	Typical failure rate in FITS
Hand-soldered joint	0.15
Machine-soldered joint	0.75
Wire wrapped joint	0.005
Fixed carbon composition resistor	0.25
Fixed carbon film resistor	0.2
Variable carbon resistor	5.0
Ceramic capacitor	0.3
Polyester foil capacitor	0.5
Aluminum electrolytic capacitor	10.0
Tantalum electrolytic capacitor	1.0
Silicon low power transistor	10.00
Silicon power transistor	100.00
LSI digital logic IC	200.00

Table 5.13 Examples of weighting factors for component failure rate

Environment		Multiplier
Normal		1.0
Airconditioned		0.5
Mobile		2.0
Seaborne		2.0
Airborne (military)		4.0
Rating		
Resistors	1⁄10 maximum watts	1.0
	½ maximum watts	1.5
	Full maximum watts	2.0
Capacitors	1⁄10 maximum voltage	1.0
	½ maximum voltage	3.0
	Full maximum voltage	6.0
Transistor/diodes	1⁄10 rated power	1.0
	½ rated power	1.5
	Full rated power	2.0
Temperature °C		
0–10		1.5
10–20		1.0
20–70		1.5

Through correct design, other factors can be brought to bear to improve the reliability of a system: for example, the addition of redundancy. This can be at the component, circuit or system level. For example, if a rectifier diode is known to fail by going open circuit, then two diodes can be wired in parallel. If it goes short circuit, then the diodes should be wired in series. Such systems are called 'parallel systems', where the reliability Rp is given by:

$$R_p = 1 - \prod_{i=1}^{n} (1 - R_{ci}) \qquad [5.12]$$

where n is the number of component units operating in parallel, each of reliability R_c.

For consumer products, there is naturally a trade-off between all these factors. Improving reliability in the production and design of equipment means increasing costs. On the other hand, the cost of maintenance (labor/downtime) and having standby equipment will decrease. This is illustrated in Figure 5.25. A range of optimum reliability should be aimed for. With the armed forces, emergency and other essential services, the cost of failure of equipment cannot be tolerated and a bias is given so that a higher cost factor is allowed.

5.7 Questions

1. List the various resistor types that you can find in the literature. Make up a table comparing the important characteristics, such as:

   (a) component range
   (b) voltage rating (DC and AC)
   (c) power rating characteristic
   (d) temperature coefficient
   (e) ageing properties.

2. Repeat the exercise in Question 1 for capacitors. Omit the power rating characteristic and replace it with loss factor.
3. What capacitor type would you select for:

   (a) a radio frequency bypass capacitor for the frequency range 50–200 MHz?
   (b) a storage capacitor for a sample and hold circuit?

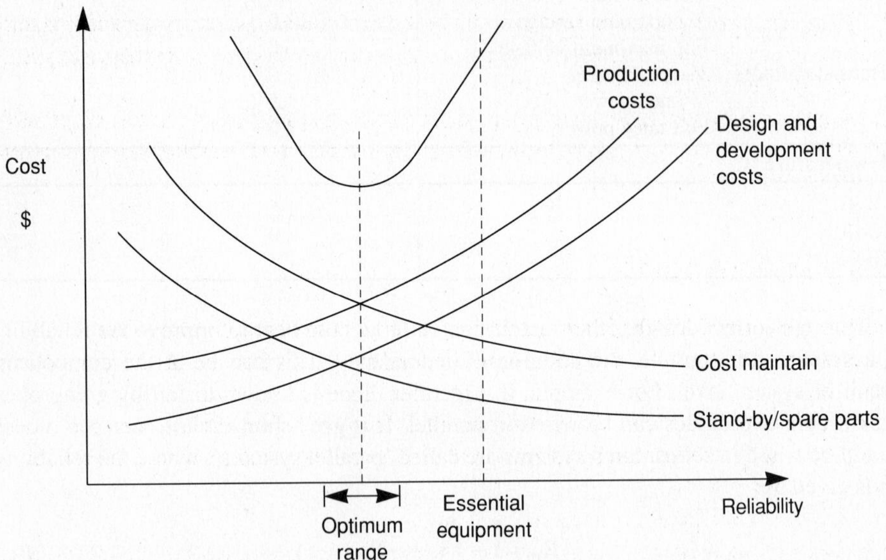

Figure 5.25 Optimum reliability point for a system

(c) a bypass capacitor for a power supply rail? If there are high frequency transient spikes on the voltage rail is a second capacitor in parallel necessary?

4. There are many types of capacitors and resistors that can be purchased directly in their final form. Many inductors, however, are wound in-house on formers that are purchased elsewhere. Why?
5. Inductors can be wound using various techniques. What is the advantage of each of the following?

   (a) layered winding
   (b) wave winding
   (c) bifilar winding.

6. What is meant by the term Litz wire and why is it used?
7. What is the difference between shielded audiocable and coaxial cable?
8. Calculate the failure rate for equipment comprising the following four component types (assume a series system):

   (a) 100,000 components of failure rate 0.5 FITS
   (b) 25,000 components of failure rates 0.3 FITS
   (c) 15,000 components of failure rates 2.0 FITS
   (d) 50,000 components of failure rates 0.7 FITS

   The equipment is expected to operate in two consecutive 8-hour shifts per day. What is the mean time between failure?
9. If the equipment used in Question 8 is mounted in a seagoing vessel which must operate in the tropics in temperatures up to 45°C, repeat the calculation. Use figures from Table 5.13. Suggest how the reliability of this equipment could be improved.
10. If the diode rectifier example given in Section 5.6 failed either open circuit or short circuit, what other components should be added to ensure adequate redundancy, allowing one diode to blow before system failure?
11. When a toroid is used to make a transformer, only complete turns are possible. However, with modern ferrite cup or core systems it is possible to obtain quarter turns. How is this possible?

5.8 References

Australian Standard 1211, 'Reliability of electronic equipment and components', (3 parts)

Australian Standard AS 1194.1 (1984), Winding wires, part 1, 'Enamelled round copper winding wire'.

Magnet Wire Pty Ltd (1973), *Magnet Winding Wire*, Hendon, South Australia.

MIL-HDBK-217E (1986), *Military Handbook, Reliability Prediction of Electronic Equipment*, US Dept of Defense, Washington.

Smith, D. J. (1972), *Reliability Engineering*, Pittman.

Surface Mount Technology Technical Design Workshop (1988), Texas Instruments (11 May), Bedford, UK.

6 Stencil and screen printing

6.1 Introduction

Screen printing was first applied to electronic circuits during the Second World War to make bomb proximity fuses. Conducting inks or pastes were silver based and resistors were made from carbon. Today the same two materials are employed in the manufacture of the low temperature polymer pastes used to print components on epoxy laminate cards (Haskard 1988).

In this chapter both stencil and screen printing will be examined, including their use in depositing solder and epoxy pastes. Alternative methods of depositing these pastes will also be considered.

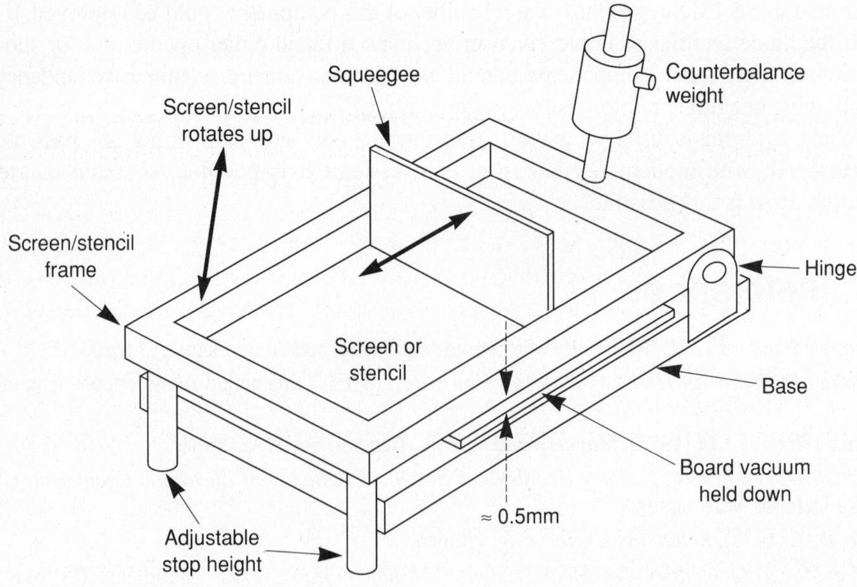

Figure 6.1 Principles of a simple screen/stencil printer

6.2 The basic process

The printing process is undertaken with a simple screen or stencil printer, as shown in Figure 6.1. The squeegee moves over the screen or stencil and forces the paste through any holes in the screen or stencil onto the board below. The pattern on the screen or stencil was previously aligned with the board so that the paste is placed in the correct position.

The board is located on a base plate and vacuum held. The screen or stencil is held in a frame that can be lowered down to be parallel with and yet separated from the board by as much as 20 mil (½ mm). Paste is placed on the screen using a stiff yet pliable squeegee and forced through the screen/stencil onto the board beneath. Surface tension between the paste and the laminated board ensures that when the squeegee has passed over and the screen/stencil has separated from the board, the paste remains on the board. The system accurately meters the amount of paste onto the board, the pattern on the screen/stencil determines the area, while the screen/stencil thickness, paste viscosity and surface tension determine, in the main, the thickness of the print.

To achieve repeatable results, the printer must have a sturdy frame, screens/stencils need to be of constant and correct tension, and the squeegee should be hydraulically or pneumatically driven.

The printer is used for two distinctly different operations. The first is the simpler, and involves the application of solder paste or epoxy paste to a board before reflow soldering. The second is more complex and involves the printing of passive components, principally resistors and conductors, although small-value capacitors can be made. These applications will be considered separately.

6.3 Solder paste/cream application

Since the application of solder paste is a relatively non-critical operation, primitive manual equipment may be used. Figure 6.2 shows a small manual stencil printer. Stencil printing methods, as opposed to syringe dispensing, provide very high throughputs.

The stencil, typically 40 mil (1 mm) thick, can be made from a range of materials such as copper, brass, phosphor bronze or stainless steel. The solder pattern to be printed is transferred to the stencil by photoengraving—that is, photolithography followed by an etch. To speed up the process the pattern may be transferred to both sides of the stencil so that etching can take place simultaneously from both sides.

Once the stencil is completed, it is glued with an epoxy to a sturdy cast aluminum or stainless steel frame. This frame sits in the printer and is correctly aligned with the laminate board pattern which is vacuum held to the base. Once aligned, this stencil is then locked in place.

Separation of the stencil from the board is important. Because the stencil is reasonably rigid, separation is slight, typically 4 mil (0.1 mm).

The squeegee is normally made of materials with a hardness of approximately 60 durometers. Typical materials are neoprene, polyurethane or Viton. With outstretched arms, the operators hold the squeegee in both hands at an angle of 45–60° and pull it towards themselves. The process is shown in Figure 6.3.

Figure 6.2 Stencil printer for dispensing solder onto a board

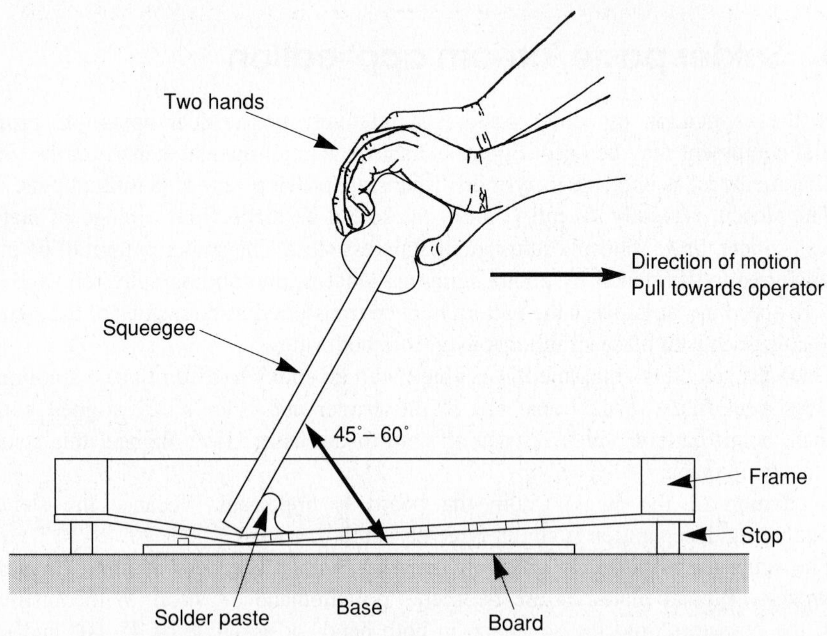

Figure 6.3 Manual operation of the squeegee

When the hinged stencil frame is raised, the printed paste can be inspected on the card. If the alignment or squeegee action is poor, the solder paste will be in the incorrect position or be non-uniform over the board. Fortunately, the paste can be washed off and the process repeated until the result is satisfactory.

Most screen printers can also be employed to dispense solder paste. The screens are woven from polyester, stainless steel or nylon and have mesh counts of about 160/inch (62/cm). The advantage of screens over stencils is that they allow printing over a much larger area. Equipment and processing using screens will be discussed in the next section. Solder cream/paste compositions have already been considered in Section 3.2.4.

6.4 Printing of polymer paste components

6.4.1 Introduction

While passive components can be added to a board in discrete form, there is also the option of printing conductors, resistors, and small capacitors. Factors such as cost and compactness of the final product decide which method should be used.

6.4.2 Screen printing

To achieve consistent printing of passive components on laminate boards, standard thick film printing equipment is employed. Figure 6.4 shows a typical unit.

The major differences to the simpler stencil printer of Figure 6.2 is that the squeegee is operated by a pneumatic/hydraulic system, the cards are loaded fully or semi-automatically, and a woven mesh screen is employed rather than an etched stencil.

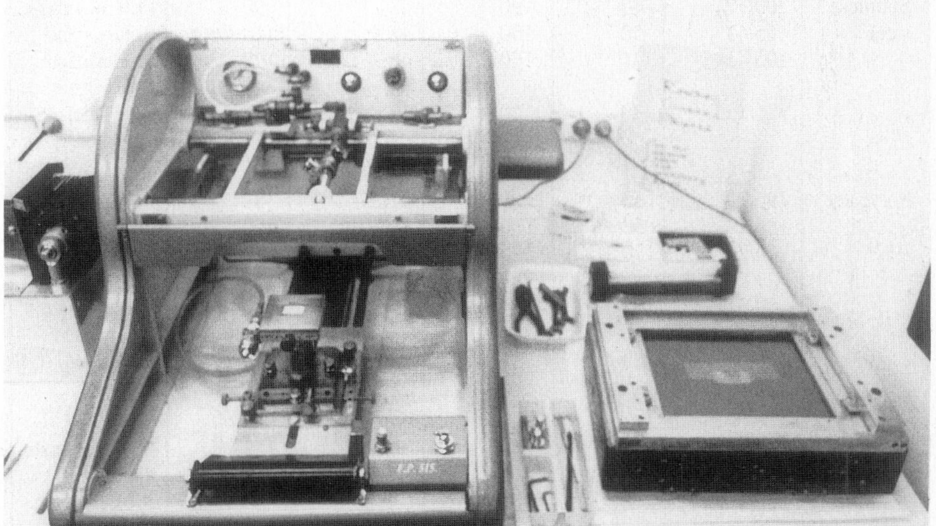

Figure 6.4 Thick film screen printing

Table 6.1 Properties of screen materials

	Mesh material		
Parameter	**Stainless steel**	**Polyester**	**Nylon**
Elasticity	Low	Medium	High
Printing properties	Good contact—small area	Good contact of board area	Can tend to stick to boards
Resilience (springback)	Low	High	Low
Thickness of printing	Low–High	Low–Medium	Low–Medium
Life	Very long	Long	Medium
Registration, definition	Excellent	Good	Satisfactory
Typical separation screen to board	25 mil (0.6 mm)	40 mil (1 mm)	60 mil (1.5 mm)

Table 6.2 Screen mesh sizes and typical applications

Material	Mesh count (inch/cm)	Fiber diameter (microns)	Mesh aperture (microns)	Percentage open area	Approx. fabric thickness (microns)	Suggested uses
Stainless steel	200/77	40	90	48	90	Polymer paste
	165/67	50	100	44.4	110	Solder paste
	105/43	71	160	48	170	Solder paste, epoxy glue
	325/125	30	50	39	67	Fine line, polymer conductor
Polyester	196/77	55	60	27.25	87	Polymer paste
	186/73	58	80	33.6	100	Polymer paste
	54/21	170	290	39.75	275	Solder paste, epoxy glue
	80/31	108	185	39.85	160	Solder paste, epoxy glue
	380/50	38	31	20.25	60	Fine line, polymer conductor
Nylon	110/43	88	128	35	145	Polymer paste, dielectrics, epoxy, glue and solder paste

Screens consist of a strong cast aluminum frame which has been machined square and across which the woven mesh is stretched to the correct tension. Important properties are the mesh material (normally nylon, polyester or stainless steel), mesh size and orientation. Of the mesh materials, stainless steel provides the highest standard of definition and registration, while polyester ensures low squeegee wear. Table 6.1 gives some material characteristics.

Mesh size is categorized primarily by the mesh count, which is the number of fibers per inch or centimeter. However, there are other important parameters such as fiber diameter, aperture and mesh thickness. Table 6.2 gives examples of screens and their uses (Hargrave 1983), while Figure 6.5 defines some of the terms.

The mesh open area O is given by:

$$O = A^2/(A + D)^2 \qquad [6.1]$$

As a general rule for mesh size selection, the paste particle size should not be greater than one third of the mesh open area.

The mesh material is stretched across the frame and fixed with an epoxy glue. Screen tension must be checked and this is normally carried out by applying a known force to the center of the screen and measuring the deflection with a dial gauge (Riemer 1986). The orientation of the mesh warp and weft to the direction of the squeegee stroke and screen frame can vary. The common angles are 0°, 45° and 22½°. Two factors are important: the strain on the fabric during printing and the resulting interference pattern from the mesh fibers on the printed paste. Maximum flexibility is given when the mesh is at 45°. This results in sawtooth edges to the printing, however. Layout of the printed artwork is normally orthogonal, so 0° orientation should provide the cleanest edges. Yet it is possible for a fiber to coincide with the artwork edge so that a narrow strip is not printed. As a compromise between 0° and 45°, orientations of 22½° are used where better definition is required.

There are three procedures for transferring the art work to a screen and defining the layers to be printed. They are the direct, indirect and direct indirect processes. With the first, the screen mesh is painted with the emulsion, making sure that it penetrates the screen and leaves no pin holes. It is leveled on both sides and allowed to dry for several

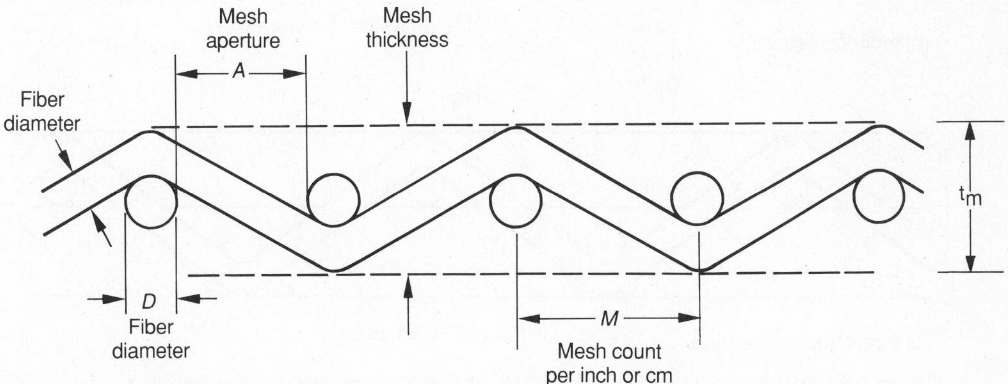

Figure 6.5 Screen construction terms

hours. The emulsion is typically a polyvinyl acetate or a polyvinyl alcohol sensitized with a dichromate solution. The screen is then exposed to ultraviolet light through a photographic positive of the contact pattern, making sure that the emulsion side of the photographic positive is in contact with the underside of the screen that will contact the substrate when printing. Those exposed sections polymerize and are made insoluble in water. The unexposed emulsion in the screen can be washed away and the screen left to dry for several hours. See Figure 6.6a. Of the three methods, this process produces the longest wearing screen and is therefore the most suitable for long production runs.

With the indirect process, the sensitized emulsion film comes attached to a clear polyester backing. The film is exposed, as described above, making sure that the photographic positive emulsion side is in contact with the polyester backing during exposure. After developing the emulsion, it is placed with the backing still in place on the underside of the screen and carefully pressed into it with a soft roller. It is then left to dry for several hours, after which the backing can be peeled off. The underside of the emulsion is never flat but follows the contours of the screen (Figure 6.6b). Nor does it

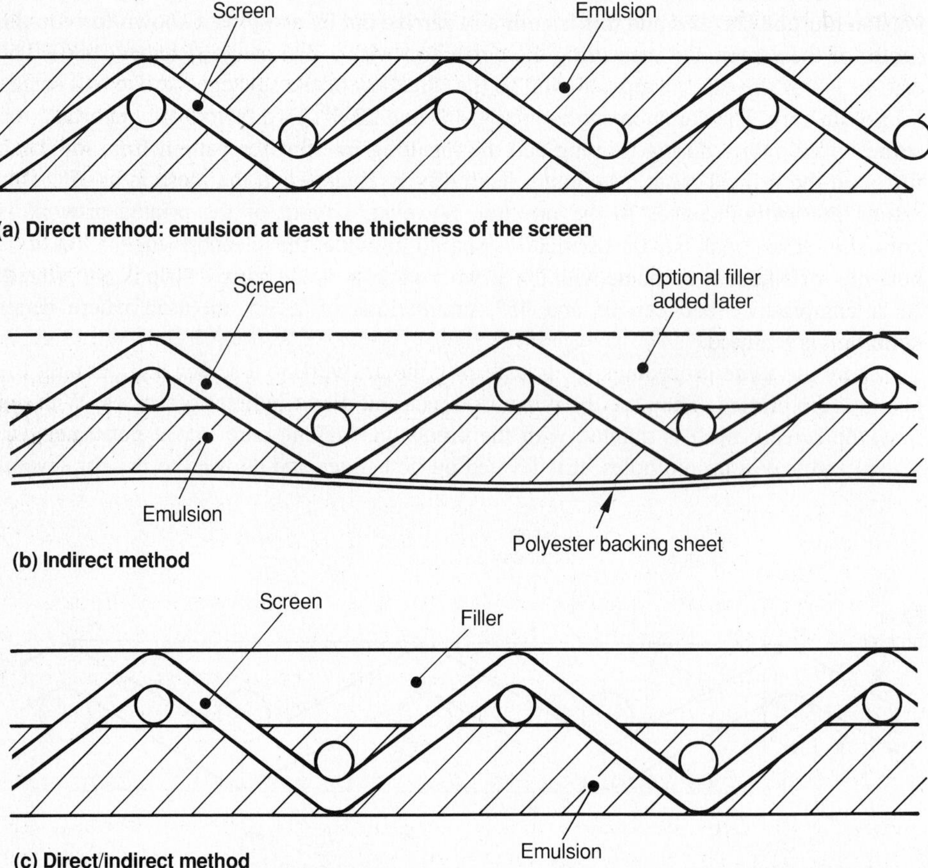

Figure 6.6 The three common methods of producing screens for printing: (a) direct, (b) indirect, and (c) direct indirect

penetrate completely through the screen. Fillers may be added to the upper side of the mesh to reduce paste wastage and simplify screen cleaning.

Although the indirect method is ideal for making screens rapidly, the emulsion and hence the printed pattern can be distorted when rolled into the screen. In the direct indirect method the unexposed emulsion with backing attached is wetted and rolled into the screen. When dry, the backing is pulled off and the emulsion exposed to ultraviolet light through the positive, as described for the direct method. Fillers are again optional (see Figure 6.6c).

As with all electronic manufacturing processes, cleanliness is essential if quality products are to be produced. Thus, screen and board cleanliness must always be maintained, both before, during and after printing.

6.4.3 Polymer pastes

Primarily designed for the printed circuit board industry, polymer pastes can be printed on any material that can withstand a firing temperature of 150°C for two or three hours (Johnson 1982*a*, 1982*b*). The pastes are thermosetting. The viscosity of the paste is controlled by either a solvent or a functional dilutant (monomer) which is often the additive that decides the paste function, whether a conductor, resistor, or dielectric type. The solvent is evaporated during curing, whereas the monomer becomes part of the polymer during cross linking. Additives to the thermoplastic pastes are in the form of finely ground particles which are held in suspension by mechanical mixing. Although infrared curing is normally used, 150°C for two to three hours, ultraviolet methods are being explored to reduce the curing times.

The advantages of these pastes are their resistance to solvents, low current noise, high wear resistance (suitable for potentiometers), and low cost. Three types of pastes are available:

1. Conductors: additives to the polymer include silver, copper and nickel, with silver being the most common. Printed conductors can be electroplated on and, with care, can sometimes even be soldered to. This is not normally necessary because the conductor can interface to the board copper track by simply overprinting the paste onto clean copper.
2. Resistors: carbon (graphite or carbon black) is used as a functional dilutant, the ohms per square depending on the amount of carbon loading. For resistors less than 1 kΩ per square, silver is also added.
3. Dielectric: no additive is needed as the polymer is an insulator, but to increase the dielectric constant, fillers such as alumina are used.

Table 6.3 gives further information on components made from these pastes. Although not discussed, the same printing system is also suitable for solder pastes and epoxy glues.

6.4.4 Layout of components

The design and layout of components to be made in polymer paste are identical to any film planar design process. For each process there is a set of minimum print widths and spacings and these must be adhered to. Figure 6.7 presents a simple set of rules, including suggested alignment patterns for the various layers.

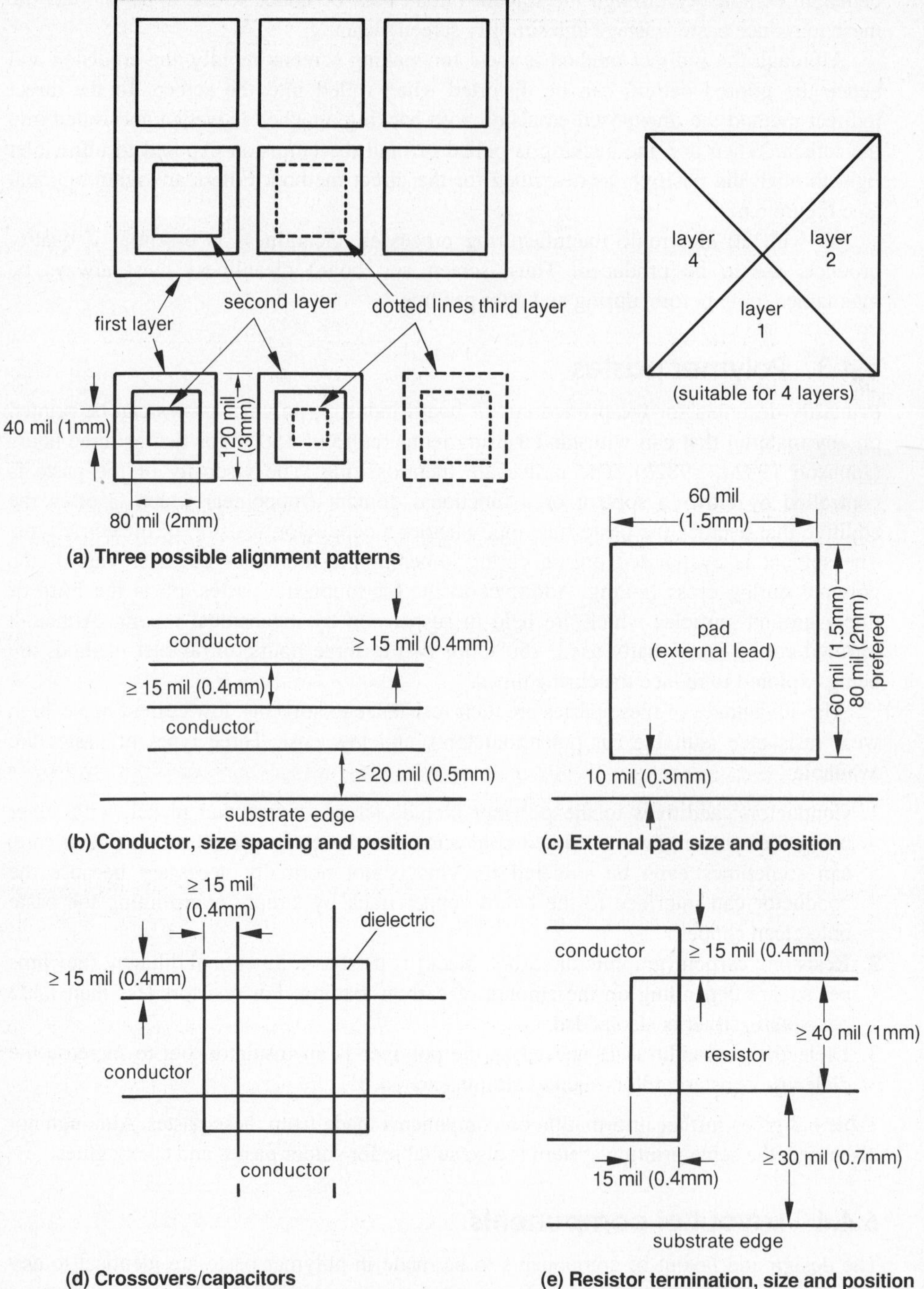

Figure 6.7 Simple thick film layout rules. Imperial units given first with metric in brackets

Table 6.3 Selective properties of polymer thick film pastes (IPC–D–859, 1987)

Conductor (silver)	
Sheet resistivity Temp. coeff. resistance	0.030 ohm/square 1800 ppm/°C
Resistors	
Range Temp. coeff. resistance Power handling Thermal stability (150°C for 1000 hours) Noise Voltage coefficient	100–100,000 ohm/square ± 250–1000 ppm/°C ≤ 10 W/sq inch (20 mW/sq mm) ≤ ± 2% change 0.01–3.0 μV/V 25–100 ppm/V
Dielectrics	
Dielectric constant (1 kHz) Insulation resistance Dissipation factor	4–6 $> 10^7$ ohm > 5

Resistor pastes are normally supplied in decade sheet resistance values—that is, assuming a constant print thickness t and bulk resistance ρ, the value of the resistance is determined by the length to width ratio, L/W. With reference to Figure 6.8a:

$$\text{resistance value } R = \frac{\rho/L}{t.W} \qquad [6.3]$$

$$= R_s\frac{L}{W} \qquad [6.4]$$

where R_s is defined as the paste sheet resistance and has units of ohm per square. To minimize area and maintain good printing tolerance, it is usual to keep:

$$⅓ \leq L/W \leq 3 \qquad [6.5]$$

Data for pastes also suggest a maximum dissipation per given area D, which is typically 2.5 W/sq inch (4 mW/sq mm) for polymer pastes. Thus for a resistor required to dissipate power P:

$$L.W = P/D \qquad [6.6]$$

Thus, using Equations 6.4 to 6.6, a paste sheet resistance can be selected and resistor size calculated. For example, a 2.7 kΩ resistor dissipating 50 milliwatt of power, using a sheet resistor paste of 1 kΩ per square with dissipation not to exceed 2.5 W/sq inch can be calculated as follows:

From Equation 6.4 we obtain $\frac{L}{W} = 2.7$

This satisfies Equation 6.5. From Equation 6.6 we obtain:

$$L.W = 0.02$$

Solving these two equations we obtain L = 232 mil and W = 86 mil. The final resistor layout is shown in Figure 6.8b. Depending on process, accuracy figures may be rounded up to give L = 240mil (19 mm) and W = 90 mil (7 mm) (L/W = 2.667 (2.714)).

Dielectric pastes can be characterized in terms of capacitance per unit area. It is usual to double print the dielectric to ensure pin holes do not cause short circuits. The dielectric layer must extend beyond the conductors as shown in Figure 6.7d.

Using laser and abrasive trimmers, screened resistors and, to a lesser extent, capacitors can be trimmed to achieve a desired value. With laminated boards, however, it is not usual to print other than non-critical components (± 20%) because when critical components are needed they are normally mounted as surface mount chips.

6.5 Epoxy glue—alternatives

With through-hole and surface mount mixed technology boards, it is usual to fix the surface mount components with an epoxy glue that will not allow them to be dislodged in the wave soldering machine. Epoxy glue can be printed, using either a stencil or screen printer, as discussed in the previous two sections. The pneumatic syringe dispenser and pin mass transfer methods are discussed below. The syringe dispenser can also be used for solder paste.

6.5.1 Pneumatic syringe dispensers

The dispensers may be simple, hand held machines, separate automatic machines, or attachments to standard pick and place component placement machines. In all cases the principle is the same. Figure 6.9 shows a typical manually operated unit that automatically dispenses a preset quantity of paste.

The pastes or glue are delivered in a syringe which is attached to the unit by a plastic tube and barrel adaptor. Various tips can be used with the syringe, including tapered, flexible, and stainless steel. The machine either accepts an industrial pressure line or generates its own compressed air.

To extrude a known amount of glue or paste there is an adjustable regulator that presets the pressure applied to the syringe and a timer that applies the pressure for a set time.

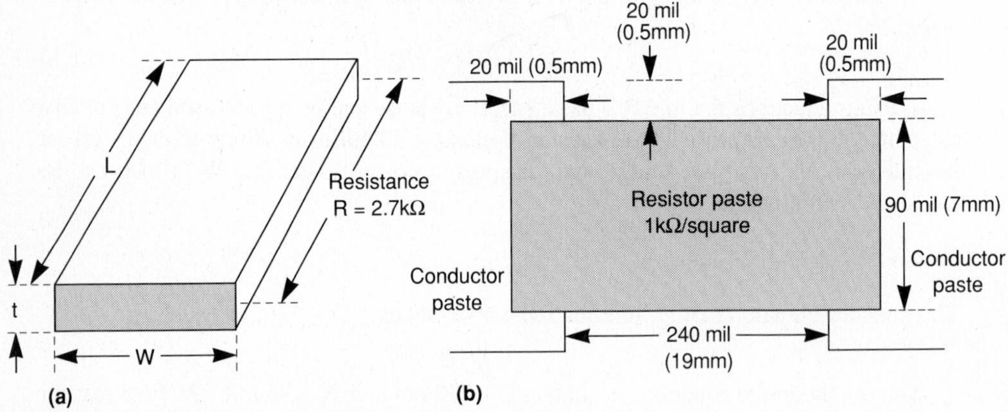

Figure 6.8 Design of a polymer paste resistor

On depressing the foot operated switch, a pulse of compressed air is applied to the syringe, forcing a measured quantity of the contents out of the tip.

6.5.2 Pin mass transfer

This is the simplest method in which a pin is dipped into the epoxy and, through surface tension, picks up a droplet. The pin and attached epoxy is transferred to and makes contact with the board so that, on removing the pin, a small drop of epoxy remains on the board. Care must be taken so that the pin does not touch the board, for this will distort the dot profile and vary the amount of glue deposited. The process is illustrated in Figure 6.10. Through using an array of pins, the process can be highly automated for volume production of boards. A special array of pins has to be manufactured for each board type.

6.6 Questions

1. List the essential components of a stencil printer, stating their functions.
2. Polyester mesh screens are considered cheaper than stainless steel and do not have a tendency to stick to boards like nylon. Why does industry not use polyester screens more?
3. Design the following resistors using the layout rules in Figure 6.7. Resistor pastes come in decade sheet resistance steps from 100 ohm to 100 kΩ per square.

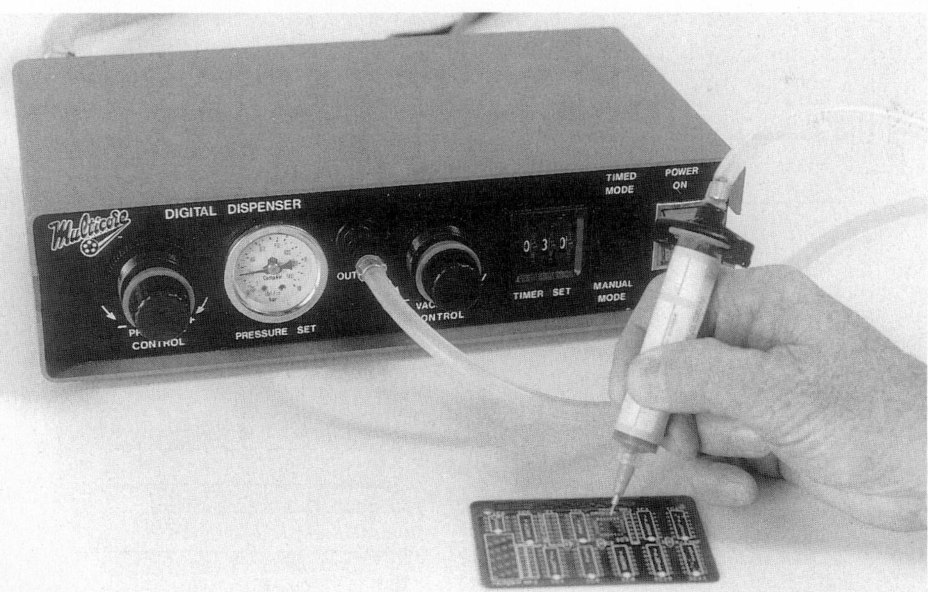

Figure 6.9 Semi-automatic liquid dispenser suitable for dispensing epoxy glue or solder pastes

Dissipation is not to exceed 2.5 W/sq inch (4 mW/sq mm). The resistors to be designed are:

(a) 10,000 ohm
(b) 560 ohm
(c) 820 kΩ
(d) 33 kΩ.

4. Three suggested alignment patterns are given in Figure 6.7. Each has differences and limitations. What are they?
5. List the circumstances that must prevail for one to print resistors rather than use surface mount chip resistors.

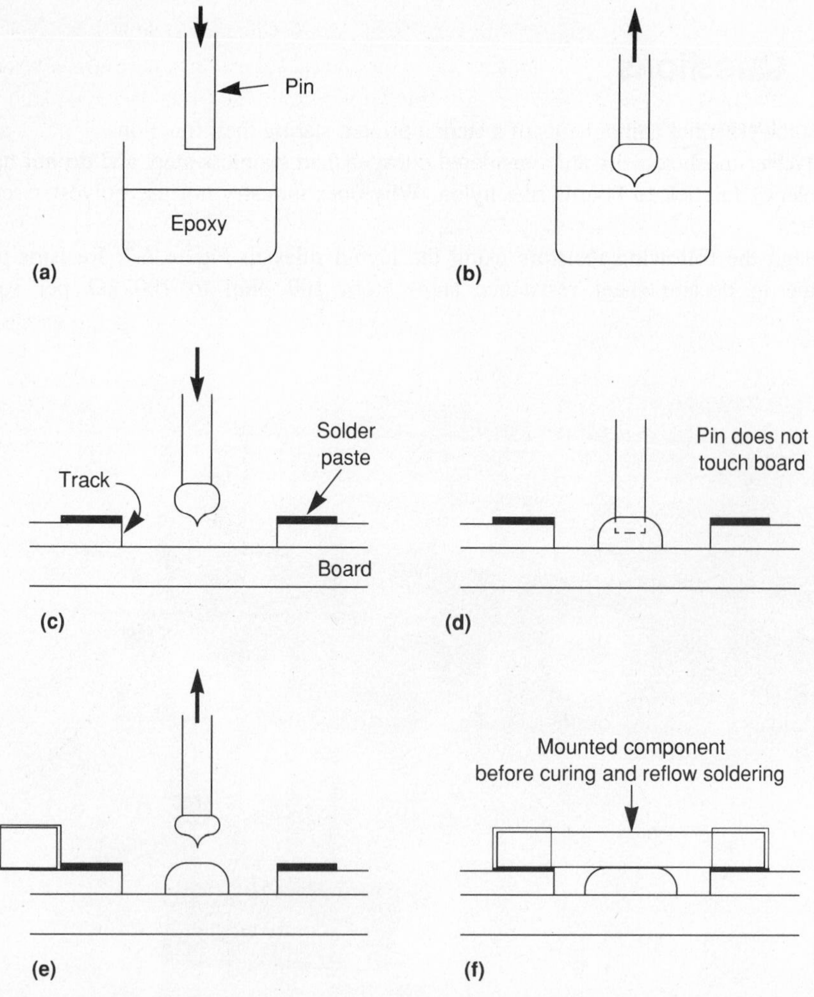

Figure 6.10 Pin mass transfer principle and an array of pins
(Philips 1986)

6. There are three processes that can be used to apply epoxy to fix surface mount components to a board. What are they? Are all suitable for both small production runs and large volume production?
7. The polymer paste layout rules (Figure 6.7) require resistor pastes to overlap a conductor by 15 mil (0.4 mm) and, likewise, dielectrics to overlap conductors by a similar amount. Why is this so?
8. Two resistors are printed using the same paste of 1 kΩ/square. The first is 40 mil by 40 mil (1 × 1 mm) and the second 200 mil by 200 mil (5 × 5 mm). What is the value of each resistor? Can you explain why they have been made different sizes?

6.7 References

Hargrave, C. E. (1983), 'Thick film screen techniques', *Hybrid Circuits*, (2) Spring, pp. 21–7.

Haskard, M. R. (1988), *Thick Film Hybrids: Manufacture and Design*, Prentice Hall, New York.

IPC–D–859 (1987), *Design Standard for Multilayer Hybrid Circuits*, Interconnecting and Packaging Electronic Circuits, Lincolnwood, Illinois.

Johnson, R. W. (1982*a*), 'Polymer thick films: Technology and materials', *Circuits Manufacturing*, **22** (7), pp. 54–60.

Johnson, R. W. (1982*b*), 'Polymer thick film applications', *Circuits Manufacturing*, **22** (9), pp. 44–50.

Mullen, J. (1984), *How to Use Surface Mount Technology*, Texas Instruments, Dallas.

Philips (1986), *Surface Mount Technology*, Philips Electronic Components and Materials, Eindhoven, Netherlands.

Riemer, D. E. (1986), 'The function and performance of stainless steel screens during the screen ink transfer process', Proceedings of *International Symposium on Microelectronics*, 6–8 October, International Society of Hybrid Microelectronics, Atlanta, Georgia, pp. 826–31.

7 Assembly of circuit boards

7.1 Introduction

The traditional printed circuit cards are epoxy laminate to which leaded components are attached through holes in the board. The bonding method is solder and it provides both the electrical and mechanical connections (Figure 7.1a). With the need to accommodate the new VLSI chip packages, which have large lead counts on small pin pitches, this method is gradually giving way to surface mount technology (Figure 7.1b). This new assembly method brings with it other advantages, including higher packing density and easier automated assembly.

Unfortunately, the changeover from one system to the other has been gradual. Not all components are available in surface mount form and the surface mount assembly equipment tends to be for high throughput or volume assembly. Further, the leaded through-hole system is a mature, well understood process. For these reasons the industry is still involved with leaded component assembly as well as surface mount, and frequently a mixture of technology is employed—that is, both are used on a single board.

In this chapter all three assembly methods will be considered: leaded through-hole, surface mount, and mixed technologies.

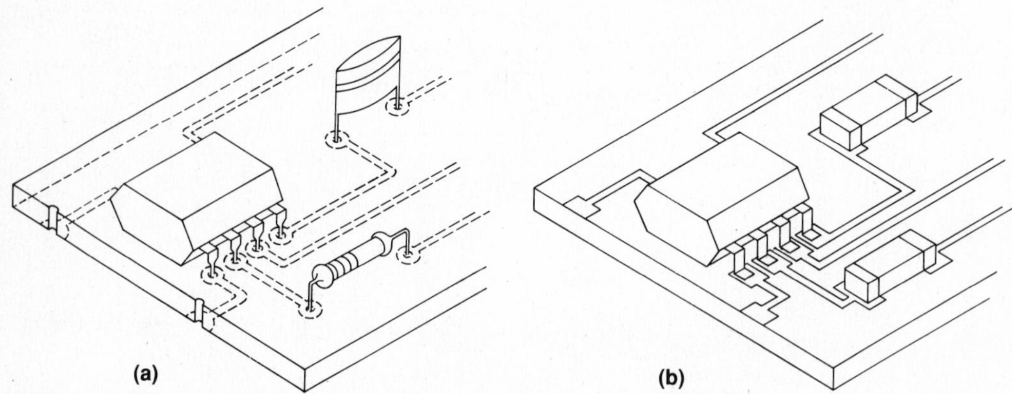

Figure 7.1 Assembly methods: (a) conventional through-hole board, and (b) surface mount

7.2 Leaded through-hole assembly

Leaded component assembly will be considered in two sections—manual and automatic. The first is normally for small volume work and prototyping undertaken on a batch basis, while the second is more suitable for high-volume production.

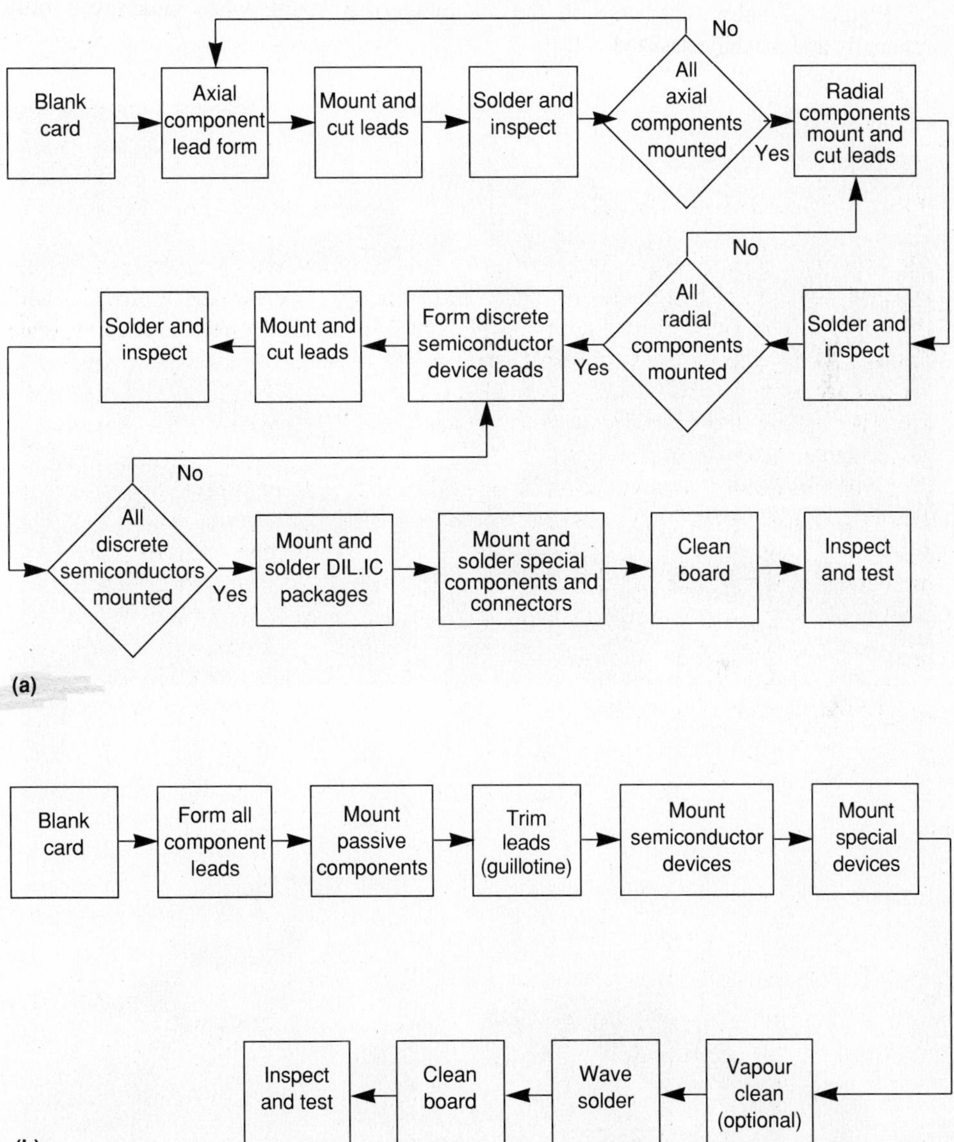

Figure 7.2 Manual assembly of a board: (a) fully manual operation; and (b) machine assisted assembly

7.2.1 Manual methods

As illustrated in Figure 7.1a, holes in the board are employed to mount the component, providing electrical connection and mechanical strength. To achieve reliable connections, the size of the hole must be related to the component lead diameter. The pad land must also be of the correct size to achieve the mechanical strength needed. All these factors must be considered when the board is designed.

Figure 7.2 shows the steps in the assembly of a board, when undertaken fully manually and machine assisted.

Figure 7.3 Component lead forming machine
(Reproduced with permission of Codan)

To assemble such a board, both axial and radial leaded components are inserted, cut and then soldered. In the fully manual case, component leads are shaped immediately before insertion into the inverted board, leads cut (if required clinched) and soldered, as discussed in Chapter 3. Many assembly lines are semimanual in that aids are employed to ensure components are assembled onto the boards in the correct position and the board is cut and soldered as a complete entity. Components are normally preformed with a preforming machine. Axial leaded components on tapes are fed into the machine and it cuts the leaded components from the bander tapes and bends them to the correct spacing. The machines can be operated manually by cranking a handle or by an electric motor. Figure 7.3 shows such a machine and illustrates the process.

Aids to ensure components are inserted into the correct position on a board (and polarity for diodes, electrolytic capacitors, etc.) may include partially assembled boards with the components to be inserted highlighted in some way, often with color marking for the benefit of the operator. Alternatively, components not to be inserted can be covered over so that only those that are visible must be placed by the operator. One of the most commonly used aids is the light projection system. The board is located in a frame on a flat panel on a bench and a computer controlled spot projection lamp placed over it. Component bins are in front of the operator, but housed under the bench top. A trap door opens, allowing particular components to be removed from the bin immediately

Figure 7.4 Spotlight system for assisting manual insertion of components in printed circuit boards. The operator is only allowed access to the correct component (Reproduced with permission of Royel)

in front of the operator. The spotlight highlights the position where the component is to be inserted into the board. For polarized components, the spot moves back and forth over the component position to highlight the particular polarity. If several of the same component are to be inserted, the trap door to the component box remains open for the appropriate time and the spotlight moves to the next location. When all these components are inserted, the trap door closes, the component bin system advances and the door opens when the next component bin is in place. The spotlight now highlights the position for the first of these new components. Figure 7.4 shows an example of a typical machine for leaded component insertion. To speed up the process, several smaller identical boards can be mounted in a frame side by side and loaded in parallel. Operators are trained to insert components with both hands simultaneously.

When all components with long leads are inserted, the board component side is backed with foam plastic sponge (made conductive to prevent electrostatic damage), so components cannot fall out, and carried in its frame to a guillotine which slices component leads to the correct length for rigid joints or slices and clinches for clinched joints. The latter is more difficult to do *en masse*, and therefore, cutting and clinching for clinched joints may occur immediately after each component is inserted. Guillotines are normally electrically operated using a foot switch.

Finally, components manufactured with short and correct lead lengths are inserted. This will include all active semiconductor devices, inductances, relays, connectors, and so forth. A light projecting system may or may not be used, depending on component numbers.

Once all components are inserted, the board is wave soldered. The printed circuit card has a solder mask so that only the pads and tracks where soldering is to occur are exposed. These exposed portions have already been tinned to assist the wave soldering

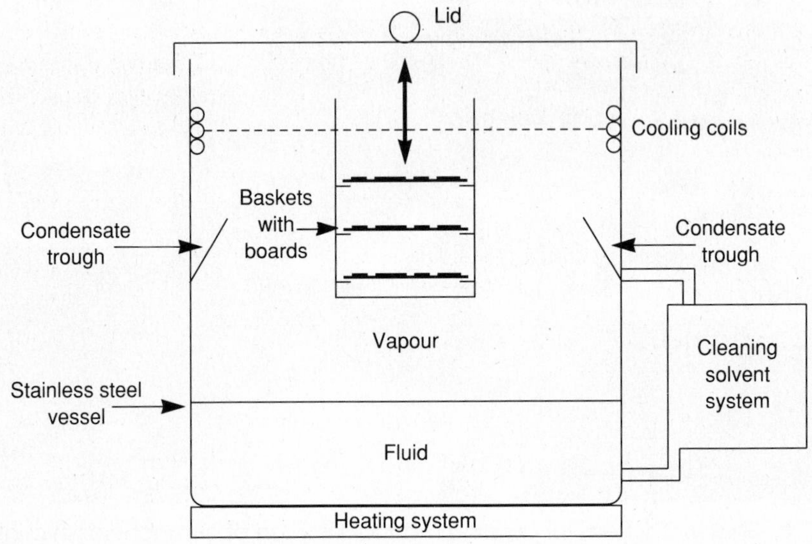

Figure 7.5 Principle of vapor cleaning

process. Component leads have also been pre-tinned by the component manufacturer. Once soldering is completed, the boards may be cleaned to remove flux and any other residue. Three methods are: ultrasonics, spray cleaning with a water-based detergent or saponifiers, and vapor degreasing. The latter is more usual when resin-based fluxes are used. Vapor degreasing is sometimes used to improve the quality and reliability of the soldering process by passing the loaded boards, prior to soldering, through the degreasing unit (see Figure 7.5). Even though considerable care may have been taken by operators, boards and components can be contaminated with substances such as greases, hand lotions, make up or hair oils.

The fluid used for cleaning is heated, producing a vapor into which the boards are loaded. The vapor is contained in the vessel by a water cooling system at the top. Condensed vapor and grease are caught in a trough and fed into a cleaning system which distills and filters before returning it to the source at the bottom of the vessel. The degreasing system, as the name infers, is only suitable for removing non-polar or grease/oil type contaminants, not polar or ionic contaminants (plating salts, etching residues, perspiration salts, etc).

Once cleaning is completed, the boards can be inspected and sent for testing.

7.2.2 Automatic methods

As mentioned earlier, the manual assembly of boards is normally undertaken on a batch basis—that is, various stages of the operation are performed in sequence on a batch of boards. While the same approach can be used for automatic assembly it is more usual to adopt a flow-line approach to achieve high-volume throughput. Figure 7.6 is a flow diagram for a typical line.

The setting up of such facilities will be discussed in Chapter 9. In this chapter we are concerned only with the equipment employed.

A pantograph machine is normally used to check the location of each component on a board. While the location can be determined directly from the board artwork, it is usual to carry out checks to allow for different board expansion and contraction rates during processing. Once this information is known, it is used to position the insertion machines.

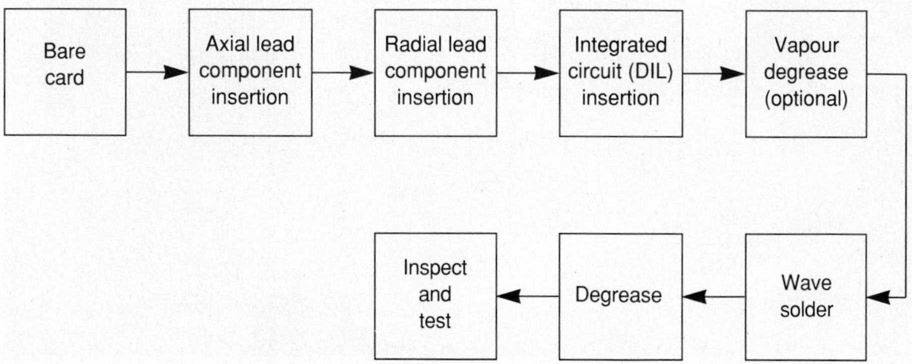

Figure 7.6 Flow diagram for an automatic leaded component assembly system

For leaded components, two types of insertion machines are required, axial and radial. Both are normally of the pantograph type, with one, or several boards moving simultaneously, to enable the heads to insert the components. Figure 7.7 shows an axial lead component insertion machine.

Components need to be fed into the insertion machine in the correct order in taped form. A component preparation machine, called a sequencing machine, accepts standard axial components in taped form and reassembles them in the order in which they are to be inserted. This includes mounting polarized components with the correct orientation. More recent machines use airflow to move components from a selection machine which cuts them in turn from reels, blows them through a common duct for lead cutting and shaping ready for insertion and crimping.

As each component is selected, it is formed, inserted into the card, guillotined and usually clinched. This final step is essential if the board has been formed by punching and all board holes are therefore much larger than the component lead diameter.

Next, integrated circuits, transistors and other special components are inserted into the board, which is then passed on for wave soldering, degreasing, inspection, and final testing.

Figure 7.7 Automatic axial lead component-insertion machine
(Reproduced with permission of Technical Components Pty Ltd)

7.3 Surface mount assembly

Compared to the insertion of leaded components, it is far easier to automate the placement of surface mount components. Consequently surface mount methods are more appropriate for high-volume, automatic production lines. It is only recently that simple professional, manual equipment has become available.

7.3.1 Manual assembly

Because all components are surface mounted, any holes in the card are for vias—that is, for connection between the layers in the card and not for component mounting. (Perhaps one exception is where the component uses the board for heat sinking.)

The steps followed for manual assembly are the application of the solder paste, placing the component and then reflow soldering. These steps are illustrated in Figure 7.8.

The solder paste may be added to the board by one of two methods: pneumatic dispensing or screening (see Chapter 6). For manual assembly, the first method is normally used. Here the solder paste is supplied in a syringe which fits into a pneumatic machine that forces a known quantity of paste out of the syringe. Both hand and foot operated controls are available. Thus the operator positions the tip of the syringe on the pad, activates the machine which, in turn, places a specific amount of solder paste on the pad. The process is continued until each pad has solder paste on it. Figure 6.9 shows such a unit. To achieve uniformity of the paste over the pad it is better to add two smaller volumes at either end of a pad rather than a larger volume in the center (Dency 1990).

Components are now assembled. In the simplest case, this is a vacuum operated pick up pen. The operator picks up the correct component with the correct orientation and mounts it on the board, the flux in the solder keeping the positioned component in place. Even though alignment of most components is not critical, since the surface tension in the molten solder pulls the component into line during reflow soldering, the process depends on the operator having a steady hand. For integrated circuits with large lead counts, positioning can be very difficult. Most organizations now use a manual pick-and-place machine similar to that illustrated in Figure 7.9. It consists of a vacuum pick-up, mounted vertically to an X–Y, rotate co-ordinated graph. When the operator brings it down on a component to pick it up, the action automatically turns on the vacuum system thereby holding the component. Placing the component on the board releases it by turning off the vacuum. The X–Y and rotate movement allows the component to be accurately positioned by the operator before placement. Some machines also have a pneumatic syringe solder paste or epoxy dispenser attachment to the side of the pick-up head so that, before placing a component, solder paste and/or epoxy may be placed onto the board.

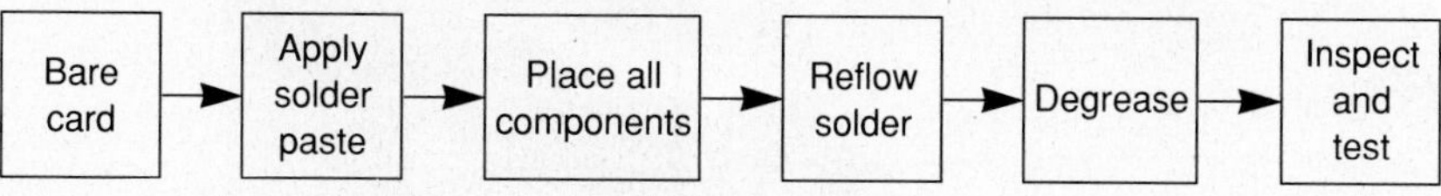

Figure 7.8 Flow line of surface mount assembly

Components can be fed to the machine in a variety of ways. Small component bins on an indexed rotating wheel, either manually or electrically turned, allow loose components to be selected. Passive and small semiconductor components can be fed from tapes on reels and larger integrated circuits from stick magazines of varying width.

Once components have been assembled, they are reflow soldered in either an infrared or vapor phase furnace, degreased to remove all flux, and then inspected.

7.3.2 Automatic assembly

Automatic assembly follows the same steps as manual assembly except that the operation is fully automated. Solder paste is applied by screen or stencil printing. Robotic pick-and-place machines, with one or more heads, are used to place components. In the simplest case, a two-headed machine may be used, the first head to apply epoxy and the second to mount the component. This is called sequential placement. There are many options, depending on the machine's capability, but they can be broken down into four categories (Siemens 1986; Philips 1986):

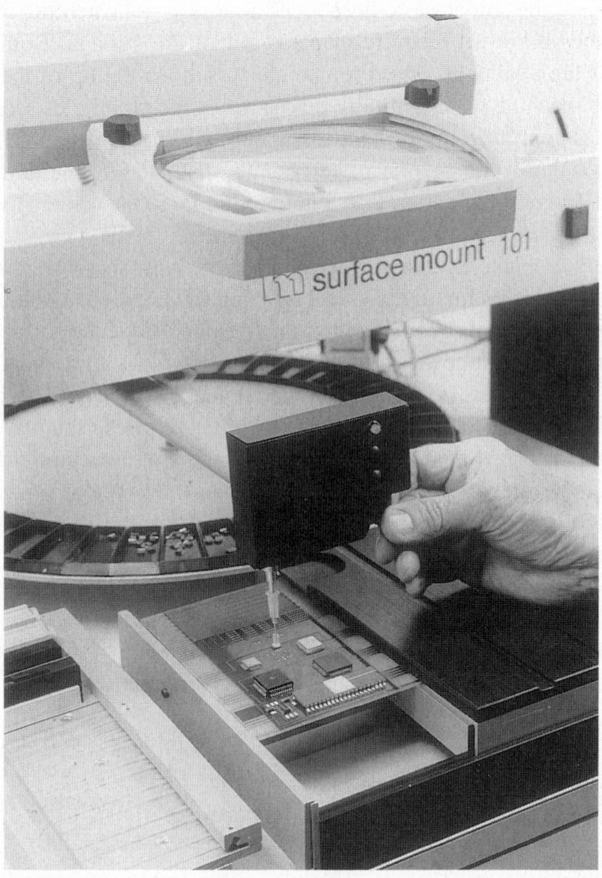

Figure 7.9 Manual pick-and-place machine

1. sequential
2. in-line
3. simultaneous
4. sequential/simultaneous.

These processes are illustrated in Figure 7.10. In sequential operation, components are placed one after the other using a single head. In an in-line operation, a machine with several heads places identical components on several boards simultaneously. It differs from simultaneous placement where the heads are used to place a group of components on a single board in one operation. Finally, the machine (or sub-machines) place various component groups on a board in the operation. Although each group of heads is operating in the simultaneous mode, each different group of components is placed sequentially.

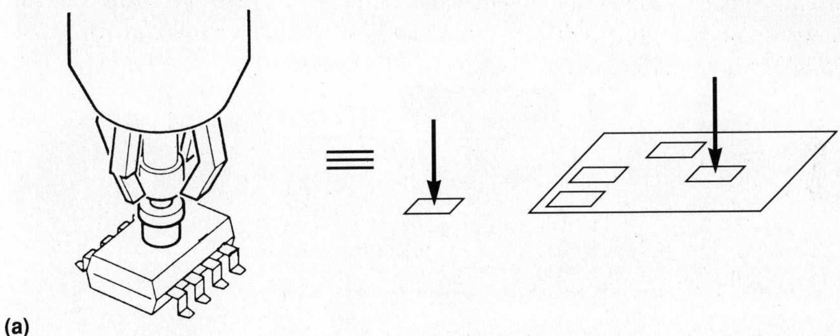

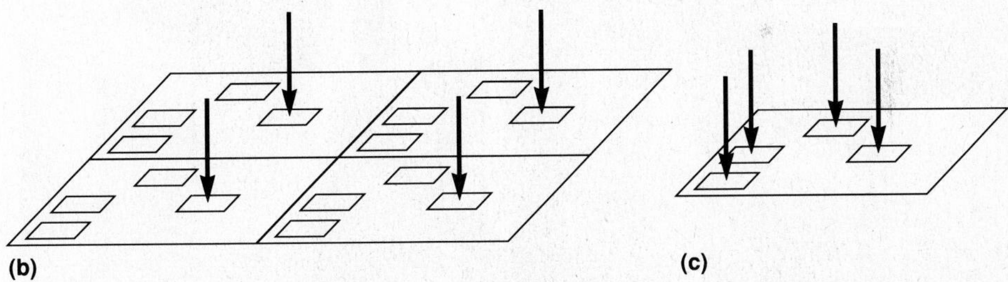

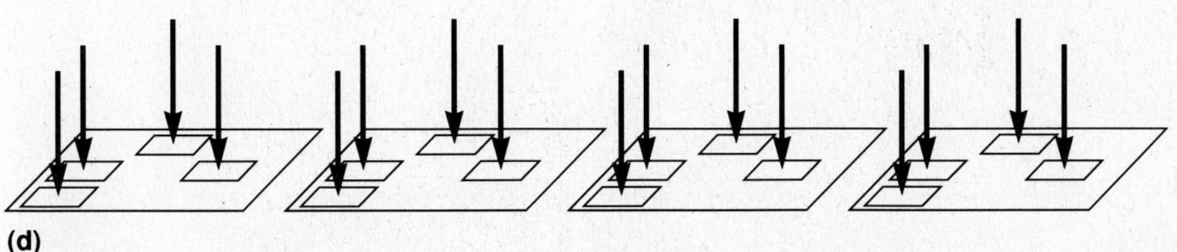

Figure 7.10 Variations in operating pick-and-place machines: (a) sequential, (b) in-line, (c) simultaneous, and (d) sequential/simultaneous

The heads used on the machines to pick up the components must be the correct type for the component being placed. For the normal flat rectangular components, the principle employed is the same—that is, a simple vacuum pick-up with mechanical grippers to position the component in the correct orientation. The head size will vary with the component dimensions. For non-rectangular components, such as a cylinder (MELF), only mechanical grippers can be used.

Packages which have either small pin pitches, or glue preventing realignment by solder surface tension during soldering, must be placed with a greater accuracy than

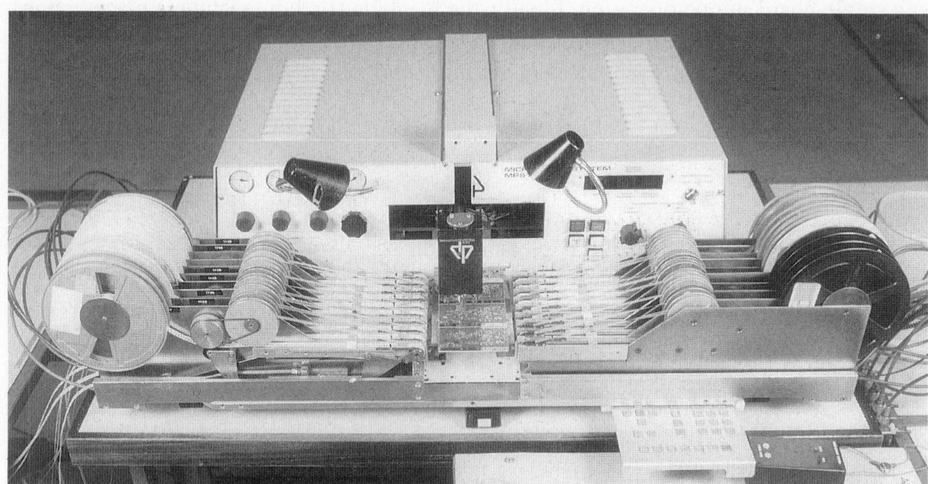

Figure 7.11 Automatic pick-and-place machine
(Reproduced with permission of Philips Components Pty Ltd)

simple mechanical positioning can provide. A pattern recognition system is employed (fiducial marks placed by each package) which provides automatic alignment.

A robotic pick-and-place machine is shown in Figure 7.11.

One of the advantages of surface mount technology is that components can be placed on both sides of a board. Loading a board therefore involves double placement and soldering. The flow diagrams for the process are shown in Figure 7.12. Epoxy glue is applied to one side to ensure that components do not fall off during the second soldering. If wave soldering is used for the second side, no solder printing is needed for that side.

The epoxy glue used to hold components can be applied using screen printing, a pneumatically operated syringe, or by pin transfer. All these methods were discussed in Section 6.4. The amount of glue delivered must not be:

- excessive so that pads are covered and soldering prevented;
- insufficient, so that the dot height does not reach the component. A dummy copper track may be added under the component to raise the dot height.

Curing depends on the type of epoxy used. Single-mix epoxy resins may be heated in an infrared oven or subjected to ultraviolet radiation for a predetermined time. Two-part epoxies, consisting of a resin and hardener, normally cure by a chemical reaction and no external heat is needed.

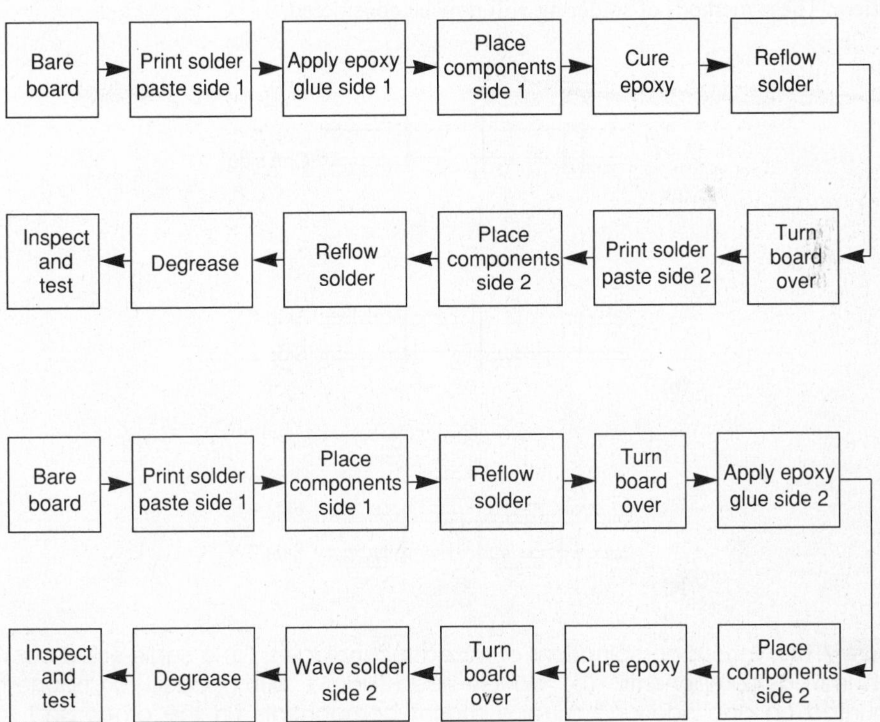

Figure 7.12 Manufacture of double-side surface mount boards

7.4 Mixed technology

In mixed technology there are several possibilities. They include:

- single-sided board with mixed components
- double-sided board with leaded components mounted on one side and surface mount components on the other
- double-sided board with mixed components on one side and surface mount on the other.

These are illustrated in Figure 7.13.

Boards of mixed technology are normally assembled automatically but manual placement can be employed if necessary. The steps to be followed are shown in Figure 7.14 (the three cases match the example given in Figure 7.13).

In the assembly examples, leaded components have been inserted last. If clinched leads are used, leaded components can be inserted before the surface mount components to be soldered onto the same side.

7.5 Soldering to posts and connectors

So far, this chapter has concentrated on the assembly of normal electronic components. In many instances, posts and connectors are used, even though wires are manually soldered to them. These methods of soldering will now be considered.

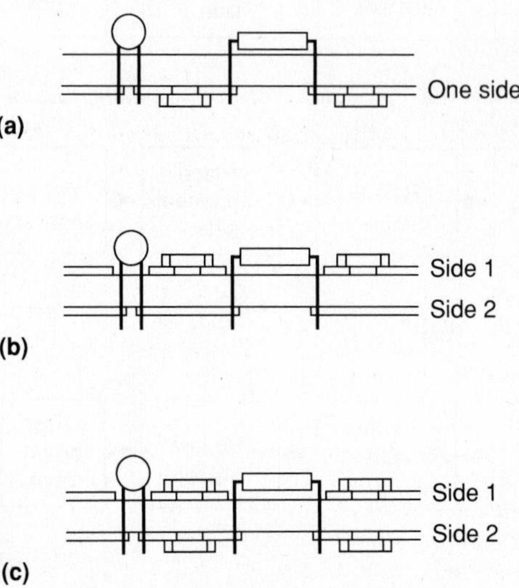

Figure 7.13 Various combinations of mixed technologies: (a) a single-sided board with mixed components, (b) double-sided board with leaded components mounted on one side and surface mount components on the other, and (c) mixed components on one side and surface mount on the other

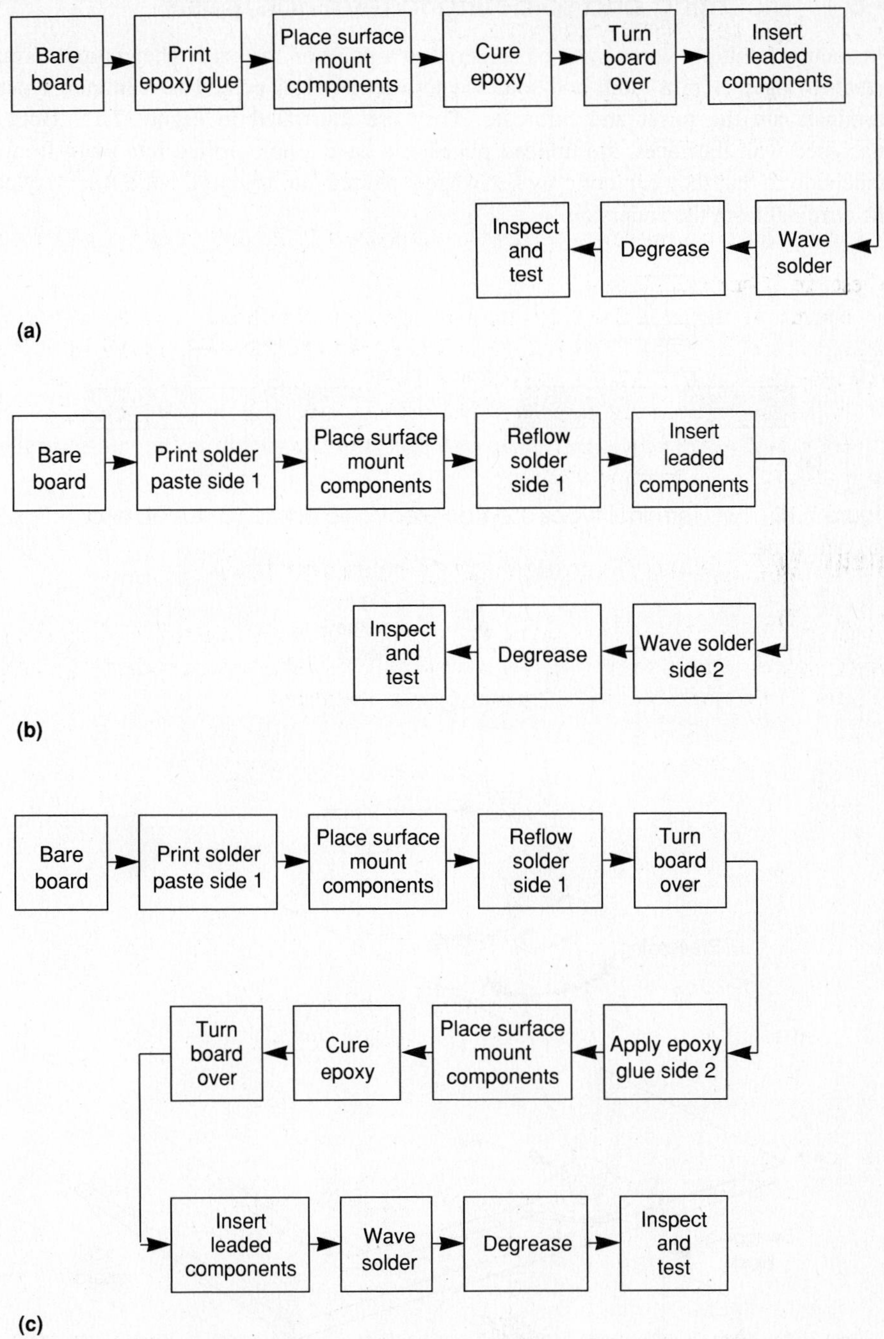

Figure 7.14 Flow diagrams for assembling the mixed technologies illustrated in Figure 7.13

7.5.1 Mounting and soldering to terminals/posts

Frequently, multiple wires must be soldered to a point on the card. They come normally bunched together in a loom and soldered to a terminal or post. Two common types of terminals are the turret and bifurcate. They are illustrated in Figure 7.15. Both are machined with flat faces, are tin/lead plated and have a hole drilled into them from the underside so that they can not only be swage mounted into the card, but a lead may enter the terminal from the underside.

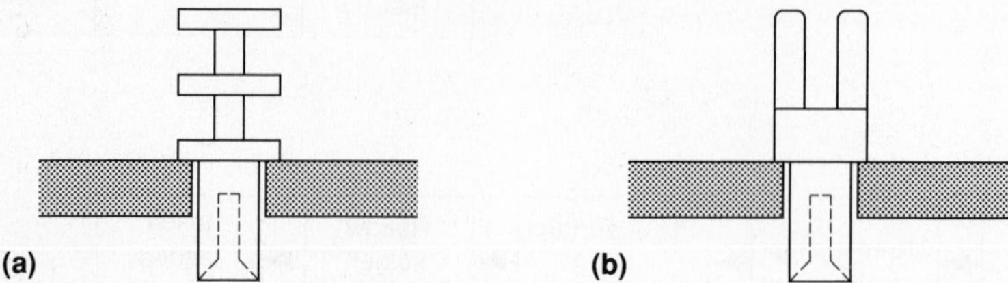

Figure 7.15 Two terminal types used on electronic cards: (a) turret, and (b) bifurcate

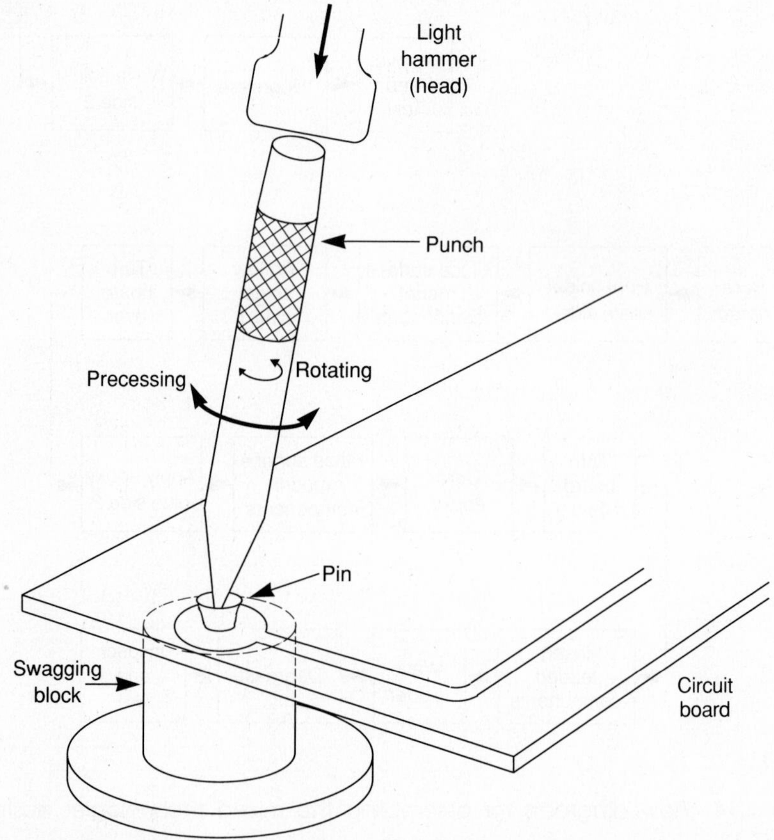

Figure 7.16 Swaging of the terminal into the card

Pins must be first mounted onto the card by swaging. A clearance hole is drilled or punched through the card to allow the terminal base to pass through. It should not be an interference fit (a typical drill size is No. 51 or 1.7 mm). For hand swaging, the pin is mounted in a swaging block, the board placed over the terminal base and pushed home hard to ensure that the terminal is correctly seated on the block. Using a light hammer and correctly tapered punch, the base is swaged; the punch simultaneously precessed and rotated in the hand and gently tapped with the hammer, as shown in Figure 7.16. Care must be taken to ensure that the base of the post is not split. When the base has been sufficiently swaged, the terminal may still rotate, but will not move in a direction perpendicular to the card.

For large production runs the pins are inserted and swaged automatically. An alternative approach is to use a knurled perimeter so that, on automatic insertion, the pin correctly grips the board.

In addition to this mechanical mounting, the terminal is also soldered to provide electrical contact and extra mechanical strength. The process is similar to forming a rigid solder joint, except that the lead is not capped. In fact, special care must be taken not to fill the center hole of the swaged terminal with solder because this will prevent the entry of a wire from underneath the card and add extra weight. Unfortunately, if the terminal is split during swaging it is almost impossible to prevent the solder from flowing into the center hole. Therefore, split terminals should always be replaced.

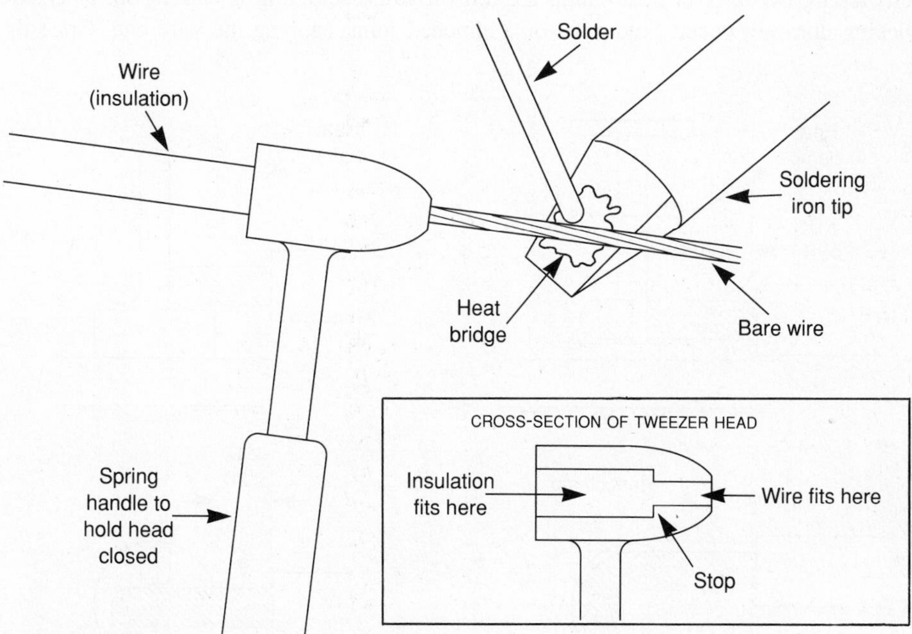

Figure 7.17 Use of antiwicking tweezers or a heat sink clamp to prevent solder running back under the insulation during tinning

Hand soldering a multistrand wire to a terminal involves two steps: first, the preparation of the wire; and second, soldering of the wire to the terminal. Antiwicking tweezers and heat sink clamps (see Figure 3.7), both of the correct size for the wire diameter, are used to ensure the insulation of the wire is not damaged. It is also important to clean all materials, that is the wire, terminal, and iron before undertaking soldering of these joints. Cleaning the finished joint is also essential.

The wire is prepared by stripping off the insulation. Mechanical strippers may be used, but thermal strippers cause less damage to the wire strands. After removing, say, ¾ inch (20 mm) of insulation (normally by twisting the insulation to keep the wire lying correctly), the stripping end is neatly trimmed with transverse end or side cutters. Should the wire strands have been disturbed in the process, they need to be layered correctly before proceeding any further. Either antiwicking tweezers or a heat sink is used to prevent solder running up under the insulation during tinning of the wire. This is illustrated in Figure 7.17.

Next, the tip of the iron is tinned, a pool of solder built up on the tip, the bare stranded wire placed into this pool and then solder fed onto the iron until the bare portion of the wire is tinned.

The soldering to the turret or bifurcate terminals is similar. In the former, the wire is soldered to one of the flat surfaces, starting with the lowest and working upwards. Four wires can be soldered to the pin shown in Figure 7.18a. The wire is shaped so that the clearance between the wire insulation and terminal is between one and two wires in diameter. The wire is formed around the pin to give 180° contact (see Figure 7.18a). The antiwicking tweezers or heat clamp are left on while soldering is carried out to ensure wicking does not occur. Solder as for a clinched joint, capping the wire end. Once the

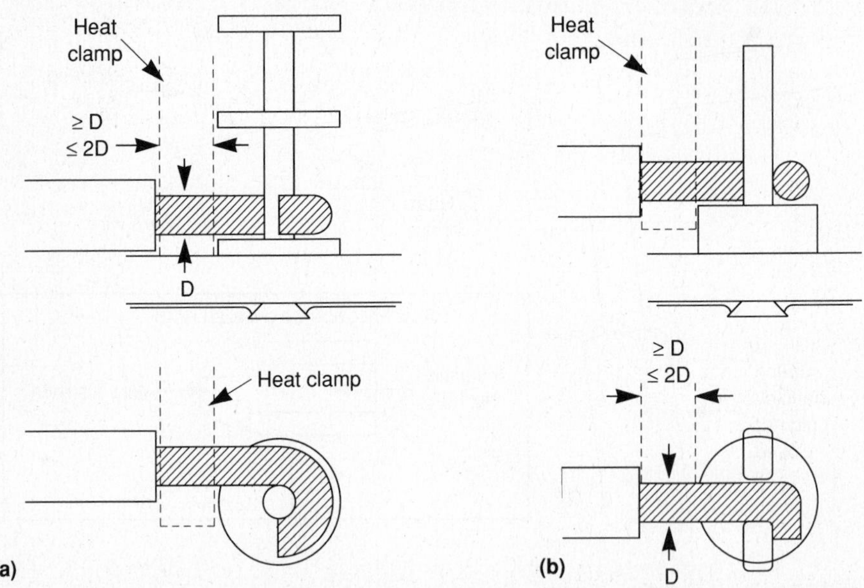

Figure 7.18 Soldering to: (a) turret, and (b) bifurcate terminals

joint has cooled, the tweezers or clamps can be removed. When a second or additional wire is soldered, care must be taken to ensure that the previous joint does not melt or wicking does not occur. It is therefore preferable to form all wires and solder them simultaneously to a terminal. Heat clamps have multiple slots to allow several wires to be clamped at the same time and provide the correct spacing.

For the bifurcate terminal, the wire is formed slightly differently, as shown in Figure 7.18b. It is simply bent at right angles with heat clamps or antiwicking tweezers used

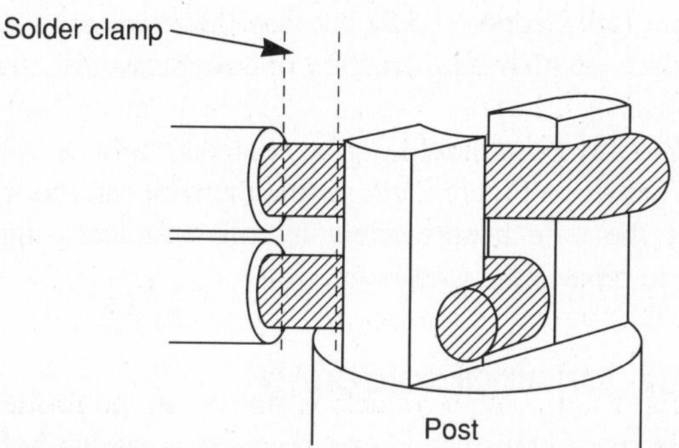

Figure 7.19 Soldering multiple wires to a bifurcate terminal

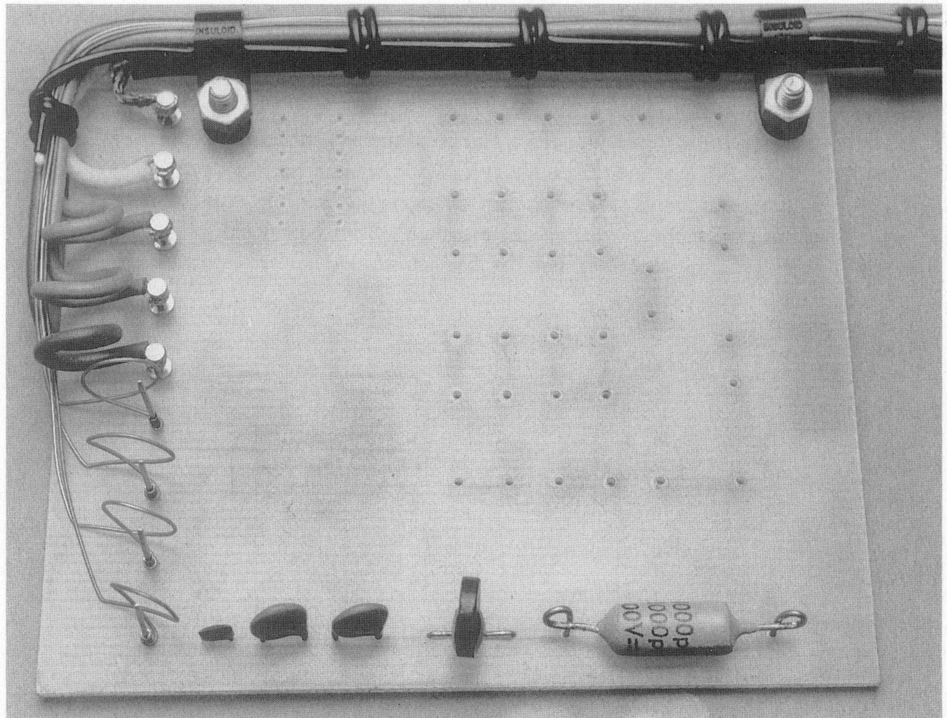

Figure 7.20 Wire soldered to posts showing the extra wire in a loop

during soldering. The joint is soldered in a similar way to a clinched joint and the cut end of the wire must also be capped. When several wires are to be soldered, alternate wires are bent in opposite directions as shown in Figure 7.19.

After completing the joint, fluxes and flux residues should be cleaned away and then the work inspected.

When soldering a wire into the bottom of a terminal, the wire is prepared in exactly the same way and, while soldering, held in antiwicking tweezers or a heat clamp. Because the hole is blind, care must be taken to ensure that flux and/or air does not become trapped at the bottom (see Section 7.5.2) because this creates a joint with poor long-term performance. Further, at high temperatures or low pressures, the trapped air can blow the joint apart.

The wire being soldered to any post should not be stretched tight. It is preferable to have a loop of extra wire, as shown in Figure 7.20, so that the wire can move freely when vibration occurs. Further, if the wire breaks, there is still sufficient length to solder another joint without having to replace the wire loom.

7.5.2 Connector cup terminal soldering

Many connectors with closely spaced pins use a cup connector, as shown in Figure 7.21a. Soldering the wire can be difficult because the hole is blind and air or flux may get trapped at the bottom, resulting in a poor joint. The wire is prepared as for the soldering of a connection to a terminal. It is then placed in the cup and, through a series of trimmings, cut to length where the distance between the end of the cup and insulation is between one and two wire diameters in length. Clean both the cup and the wire.

Soldering can be undertaken with a hand iron or a pair of resistance heating tweezers. With the hand iron, heat the cup at the back, getting as little solder as possible onto the outside of the cup. Fill the cup with solder and continue to heat until all flux and air have been removed from the cup. At this stage it has been tinned correctly. Allow it to cool, clean away the flux and inspect. If resistance tweezers are used, the process is the same. The cup is gripped with the tweezers and power applied with the foot switch. Do not

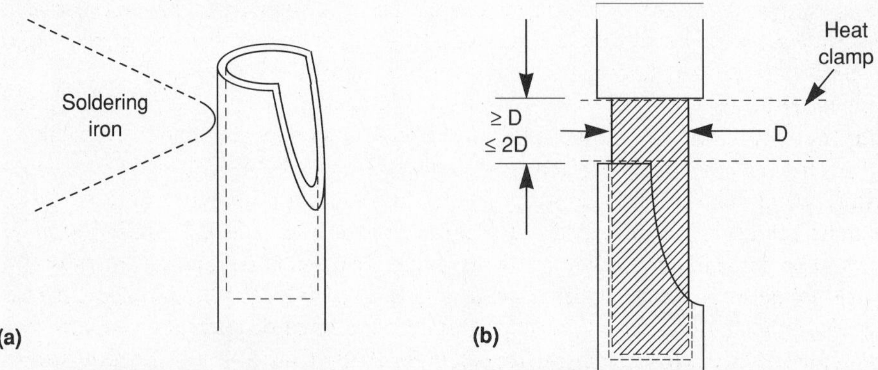

Figure 7.21 (a) A cup terminal, and (b) final connection

apply power until the tweezers are in position because arcing can damage the terminal. Feed the solder into the cup.

To make the wire connection, the cup is again heated and the wire held in the antiwicking tweezers or heat clamp as it is lowered into the cup. Rock it slightly to ensure it is positioned correctly and there is no trapped air or flux. Remove the hand iron or resistance tweezer power to allow the cup to cool. Be careful not to move the wire. Clean and inspect the joint.

7.6 Health and safety

Safe practices must be used when performing soldering, cleaning and other assembly operations. Safety standards and procedures must always be followed. There are two areas of concern: chemical and physical.

7.6.1 Chemical hazards

Many of the cleaning solvents, acids, alkaline baths, and etch solutions, are dangerous to health and property. In the latter case, there is the risk of corrosion, staining, and flammability. Consequently, correct construction materials should be used, good ventilation installed, safe areas set aside for bulk chemical material storage, and all areas fitted with correct signs and fire extinguishers.

Dangerous chemicals can be classified into six categories, depending on whether the hazard arises through ingestion, absorption, inhalation, contact, flammability, or water reactivity. The eyes and exposed skin are particularly vulnerable and appropriate protective gear such as glasses, face shields, gloves, aprons and shoes (all made of chemically resistant material) should be worn by operators. Arm operated showers for washing exposed skin should an accident occur, established first aid training and procedures, clearly displayed ambulance, hospital and doctor telephone numbers are all essential to a safe working environment. Breathing apparatus should be readily available for rescue operations. Chemical areas should be well ventilated, and smoking and eating prohibited.

Early symptoms of chemically related illness may be skin irritations, sore throats, headaches, dizziness, tiredness and/or loss of appetite. Medical advice should be sought to correct the immediate problem and to prevent long-term effects occurring.

Some chemicals are also environmentally unfriendly. Since 1987, the Montreal Protocol has controlled the manufacture and use of substances that deplete the earth's ozone layer. The electronics industry is responsible for about 12% of the ongoing depletion, mainly through the use of CFC-113 as a solvent for flux removal (Lea 1991). Since most halogenated hydrocarbons deplete the ozone layer, causing global warming, they are to be gradually phased out of the industry. Trichloroethane, tetrachloroethylene and dichloromethane are now being used instead of CFC-113. New plant, such as explosion proof boiling isopropyl alcohol cleaners, are being developed as well as new materials such as the semi-aqueous solvents (Hayes 1989). Consequently, the next decade will see a rapid change in technologies to eliminate those materials that are environmentally unfriendly.

7.6.2 Physical hazards

The temperature of molten solder is such that, on skin contact, severe burning occurs. Eye damage is a special concern when undertaking mass soldering processes, such as wave soldering because any moisture present will cause solder splashing. Protective glasses and face shields, protective clothing, heat resistant gloves and aprons are essential. Signs indicating danger and hot solder should be displayed.

The practice of flicking soldering irons to remove excess solder and trimming of leads so that the cut section becomes a projectile must be prohibited.

In addition to these special hazards in the electronics industry, precautions to ensure basic work safety must still be taken. These include care in lifting heavy materials, protection of presses, and establishing satisfactory ergonomics in the work area.

7.7 Questions

1. You are required to advise on the setting up of a surface mount assembly line that can handle boards at the rate of 10 per hour. What equipment would you suggest is needed?
2. The flow diagrams for mixed assemblies of boards (see Figure 7.14) leave leaded component assembly until last. Modify these flow diagrams to place the leaded components earlier in the process. Assume all leads are clinched.
3. Explain the operation of a vapor degreasing unit.
4. For large volume assembly of surface mount components, a bank of pick-and-place machines, each with four heads, can be arranged to operate in a variety of ways, including sequentially, simultaneously, and in-line. Propose a number of configurations for setting up these machines. You are to assume a particular job where there are four printed circuit cards to a board, each card requiring 130 surface mount components, to be placed on one side only, and that a pick-and-place head, on average, takes 2 seconds to place a component. Estimate the card throughput for each of the configurations you have proposed.
5. Make a list of the chemicals used in the electronics industry. Find out which ones constitute a health hazard and what precautions should be taken.
6. Many of the chemicals used in the electronics industry are environmentally unfriendly. List these chemicals. Are there alternative chemicals or water-based systems that can be used instead? Make a second list of these alternative approaches.
7. Fiberglass boards passing through an infrared reflow furnace exit with small components well soldered, but the larger components, such as 48 pin J lead packages, have some leads where the solder joint is granular, of dark gray mat color and a high resistance, possibly open circuit. What steps would you initiate to investigate and solve this problem?
8. List the operator hazards that can occur if a wave soldering machine is incorrectly set up and operated. What safety precautions must be taken to protect operators?
9. The wearing of safety glasses is essential for many operations in the manufacture and loading of boards. List those operations and the hazard if the glasses are not worn.

7.8 References

Dency, G. C. (1990), 'Solder addition in automated assembly', *International Journal for Hybrid Microelectronics*, **13**(1), pp. 12–21.

Hayes, M. E. (1989), 'Semi-aqueous assembly cleaning: the best choice for SMT cleaning', *Proceedings of the Conference on Defeating the Ozone Problem*, November, Teddington, United Kingdom.

Lea, C. (1991), 'Solvent alternatives for the 1990s', *Electronics and Communication Engineering Journal*, **3**(2), pp. 53–62.

Philips (1986), *SMD Technology*, Philips Electronic Components and Materials, Eindhoven, Netherlands.

Siemens (1986), *SMD Automatic Placement Systems*, Siemens AG, Munich, MS-72, HS-180.

8 Inspection and testing of cards

8.1 Introduction

Since an organization is only as good as the products it produces, quality is essential. While various standards and specifications can be used to test or police for quality (AS3900 to 3904, ISO–9000 to 9004, NATO–AWAP-1, MIL-Q-9858A, British DS 05/21), maintaining the quality of a product requires a commitment by *all* company employees. Quality is an attitude: the directors must be committed and prepared to invest money to set up adequate facilities; factory workers must take a pride in their work; and aftersales people should be helpful in following up any problems. Within organizations there is often a quality department (or, in smaller organizations, a single person) responsible for establishing procedures to ascertain and maintain quality (including the preparation of manuals). This department or person must be seen to be working with and assisting all other sections to obtain a company objective, namely product quality (Pyzdek 1989).

Most standard documents define quality as fitness for purpose. In one sense quality is never achieved because a product can always be improved. In reality, there is a price/quality trade-off, with the customer accepting the responsibility for that trade-off. However, one thing is certain: if the whole of an organization is committed to achieving quality products the first time, then the defect and rework rate will be low. This will create a more efficient design and manufacturing system, resulting in lower cost for quality.

Quality will be achieved only if integrated into the design and manufacturing process. There should be provision for inspections and tests, and the results should be fed back to improve the design and manufacturing stages. As a rule, faulty components/designs should be identified as early as possible because if they are left to be detected at the next level of test, the correction cost is generally higher (Siemens 1986). Consequently, the level of testing should follow this order:

1. electronic components
2. bare cards
3. assembled cards
4. working cards
5. burn in
6. life tests.

To minimize the testing of incoming components, quality assurance methods, based on statistical sampling, are normally employed. Approval may be given to suppliers who have undergone an earlier inspection or examination to verify that they can make components to a required quality. Their components need not undergo further quality assurance tests. In all other cases, however, tests are essential if quality products are to be produced (Richards & Footner 1987).

In this chapter we will not consider the testing of simple electronic components, but concentrate, by way of example, on the testing of printed circuit boards at various stages of manufacture and assembly.

8.2 Board faults

The amount of testing involved in the manufacture of a product should be carefully watched and controlled. It must be sufficiently thorough to pick up all faults, but to go beyond this will mean that testing stations become bottlenecks in the manufacturing process or excessive capital and resources need to be spent on testing equipment (Siemens 1986). Naturally, some types of equipment, such as medical and military, may require extensive testing with customers willing to pay extra to ensure that the reliability they require is achieved.

Some prior knowledge of the types of faults which can occur in making and assembling boards can also help to reduce the numbers of tests that need to be undertaken, thereby reducing costs. Various sources (Siemens 1986; Teraoka 1986) suggest the percentage rates of certain faults as shown in Figure 8.1.

With wave soldering, most of the faults are simple short circuits (such as bridges)

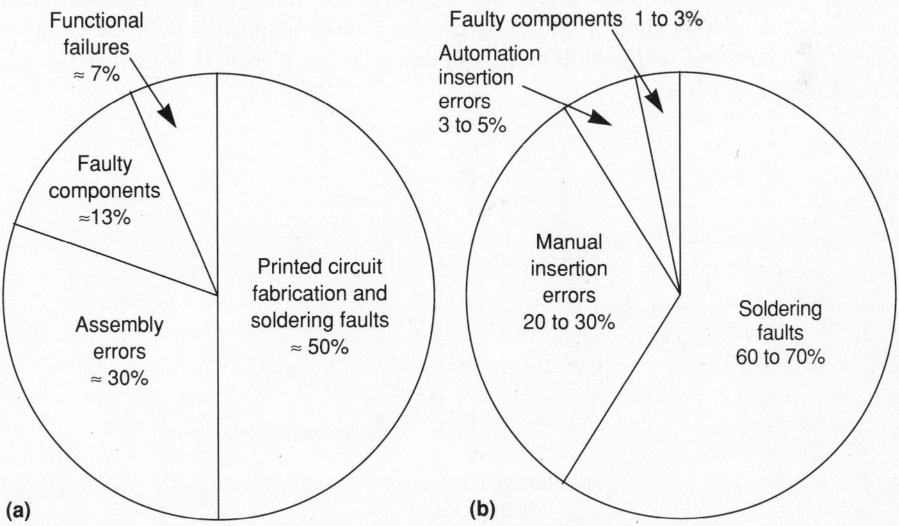

Figure 8.1 Typical percentage faults on printed circuit boards in (a) Europe, and (b) Japan

while, with surface mount components, reflow soldering of the joints tends to cause high resistance. Bearing these in mind, it is important that inspection and testing concentrate on the board fabrication and soldering.

A problem common to all boards is the wetting of component leads. Should this be a difficulty (either surface mounted or leaded components), tests should be carried out to determine the 'wettability' of those leads. The International Electrotechnical Commission (IEC) specifies procedures for such testing (Documents 68.2.54 and 68.2.20). It involves dipping the leads into molten solder and measuring with a load cell the vertical component of the surface tension force, as shown in Figure 8.2. Such measurements are important because it has been discovered, particularly in the case of surface mount components, that solderability deteriorates with time—even after only a few months.

8.3 Board inspection

Until recently, boards were inspected by operators. This included bare boards, which were checked specifically for track open circuit and short circuit faults. The inspection of assembled boards was more difficult because components had to be present, of the correct type and polarity, undamaged and, in the case of surface mount, correctly aligned. All solder joints had to be acceptable with no joints missed. Fluxes and all other contaminants also had to be removed.

Computer assisted inspection is now the norm with the introduction of full computer inspection fast approaching. In the first instance, an X–Y table allows the board to be automatically scanned. A zoom stereo optical projection system is used to view the board. Forty-five degree angle heads are employed to scan around the surface mounted integrated circuits to view their position accuracy and soldering. Alternatively, a video viewing system can be used where the signal is fed through an image enhancing microprocessor system (Juha 1987). This process can be automated with the computer comparing the images with inbuilt models of defect joints. If there is no match, the joint is considered acceptable.

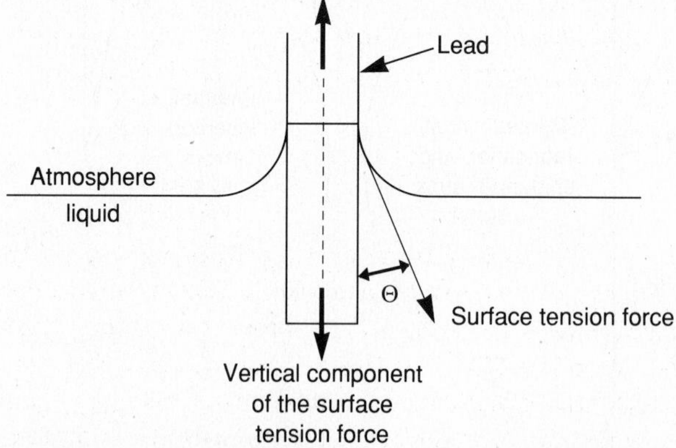

Figure 8.2 Measurement of solderability

Other means of automatic inspection of soldered joints are also available (Lea, Howie & Seah 1985). One is a piece of equipment similar to the closed loop laser soldering system which uses a YAG laser of 1.06 microns wavelength, with a spot size of 17 micron diameter to heat each joint a few degrees in temperature. Power is applied for about 30 milliseconds. An infrared sensor picks up the heat re-radiated by the joint, each joint having its own thermal signature. If there is a reduced thermal mass, then the joint has a void, blow hole, insufficient filling, or some other defect. Should the wrong surface absorption occur, due to a dull or granular finish, cold joint, contaminated solder or poorly removed flux, the joint can also be classified as unacceptable. The unit can examine joints at the rate of about 10 per minute with about 80% accuracy (this is as good as a human operator).

Another alternative is a machine that examines the vibrational characteristics of a joint. If a joint is stimulated, say by a jet of compressed air, the resonant frequency of vibration is determined by the lead material and geometry. Because of geometry constraints, a soldered joint will have a higher resonant frequency than an unsoldered or poorly soldered joint. Typical frequencies for soldered joints of integrated circuit packages leads are 85 kHz and upwards, while unsoldered joints are typically in the region of 20–50 kHz (Keely 1989). A laser Doppler vibrometer is used to measure the velocity of the vibrating lead stimulated by the direct jet of compressed air. The peaks in the frequency spectrum indicate whether the joint is soldered or not. The machine cannot recognize all faults. It detects when no mechanical joint has been formed—that is, those defects that are open circuit, such as dewetting, wicking, non-wetting and lifted leads. Shorts, bridges, solder balls or splashes, cold joints, insufficient or excess solder are seen as good joints. Cracked joints may be detected in some instances.

8.4 Board testing

Automatic testing has been used for large volumes of board production since the 1960s. With the introduction of surface mount techniques, board packing densities have increased significantly so that, even for moderate board quantities, automatic board testing is essential. A further advantage of such machines is that they allow the collection of data on faults which, in turn, allows isolation of problems in component types, design and assembly areas. The testing can be undertaken at three levels:

1. bare board
2. in circuit
3. functional.

The testers can be either dedicated to a particular board type or universal (accommodating a range of board styles and types). Most are based on the bed-of-nails concept, which consists of a spring loaded pin brought down on each test point on the board with a pressure of 100–200 g, to ensure a good contact. An array of such pins is called a 'bed-of-nails' and allows simultaneous contact with all test points on the board and signals to be applied. Up to several tens of thousands of pins can be accommodated. Figure 8.3 is an example of such a unit while Figure 8.4 indicates a number of different pin head arrangements and their typical application.

The boards must have test points. For leaded through-hole boards, it is usual to use the plated-through holes. A tester is provided with all the hole information from the computer aided design tape, but only pins for the required holes are actually set up. With surface mount boards the pins must not be applied directly to the ends of the components, because the 100 g pressure can cause a faulty joint to appear acceptable. Further, some ceramic components can be damaged by the pins. On surface mount boards vias can be

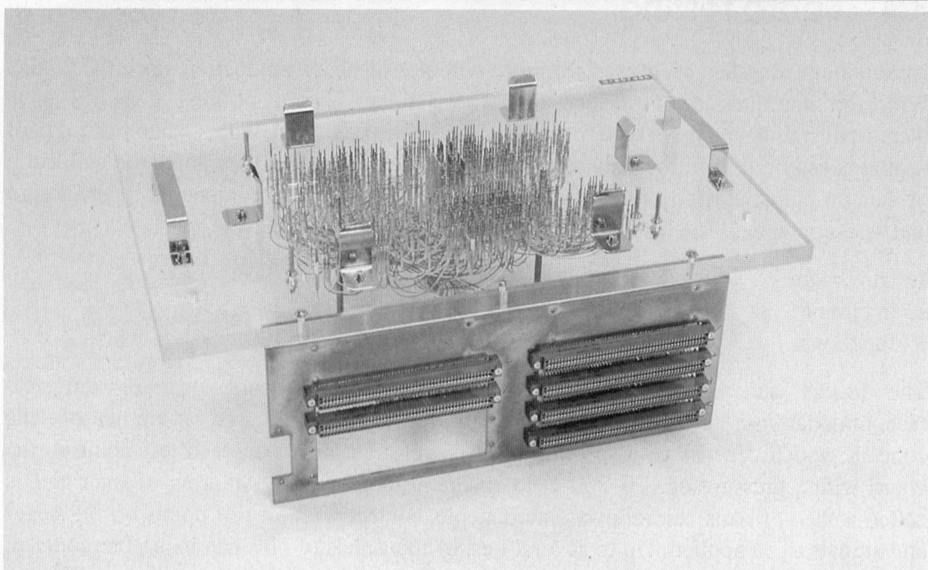

Figure 8.3 Bed-of-nails board tester
(Reproduced with permission of Okano)

used as test points and, where additional ones are required, special small test pads are added, their diameter dependent upon the bed-of-nails grid.

The normal grid for leaded component boards and the preferred grid for surface mount boards is 0.1 inch (2.5 mm). Here test pads should be greater than or equal to 0.05 inch (1.3 mm). For grids of 0.04 inch (1 mm) test pads should be greater than 0.03 inch (0.7 mm) (Beselin & Schauflinger 1987). Where grids are finer than 0.1 inch (2.5 mm), the test pins are smaller and more fragile so tend to become easily damaged. Since pins are expensive to replace, the 0.1 inch grid should be used whenever possible. It is also preferable to have all test points on one side of a board. While it is possible to have bed-of-nail pins on both sides simultaneously, it is a far more difficult exercise and the equipment is considerably more expensive to purchase and maintain. All test points should be on the soldered side of a board and soldered. Should there be a tall component, test points should be separated from it by at least 0.2 inch (5 mm). Figure 8.5 shows typical test points on a surface mount board.

PLUNGER	TIP STYLE	APPLICATION COMMENTS	PLUNGER	TIP STYLE	APPLICATION COMMENTS
A (or) G	Concave	Long Leads, Terminals, and Wire Wrap Posts.	LM	Star	Plated through holes, Lands, Pads—Self-cleaning.
B	Spear Point	Lands, Pads or Plated through holes.	T (or) K	3 or 4 Sided Chisel	Plated through holes—Cuts through contamination.
C (or) F	Flat	Gold Edge Fingers—No marks or indentations.	U	Crown .040	Lands, Pads, Leads, Holes—Self-cleaning.
D (or) J	Spherical Radius	Gold Edge Fingers—No marks or indentations.	V	Crown .060	Lands, Pads, Leads—Self-cleaning.
E	Convex	Plated through holes.	W	Crown .050	Lands, Pads, Leads—Self-cleaning.
FX	Flex Probe	Contaminated Boards or Conformal Coating.	X	Tapered Crown	Lands, Pads, Leads, Holes—Self-cleaning.
H	Serrated	Lands, Pads, Leads, Terminals.	Y	Tulip	Self-cleaning Leads, Wire Wrapped Terminals.

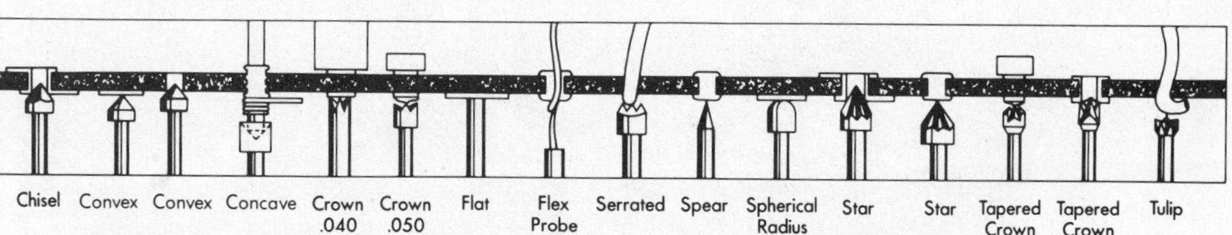

Figure 8.4 Examples of different pin heads and their application

The problem with the bed-of-nails approach is the force needed to keep all of the pins in good contact with the test points on the board. If there are 10,000 test points and each pin requires a 100 g pressure, the total force to be exerted is 1000 kilograms or a tonne. A new type of tester has been devised (Conti 1986), consisting of two computer controlled probes. They can come within 20 mil (0.5 mm) of each other and test points at the rate of 350 per minute.

8.4.1 Bare board testing

Bare board testing checks for short circuits between tracks and continuity of tracks. (The testing of insulation and dielectric strength must be carried out in other test machines.) The bed-of-nails makes the contacts and, under computer control, a current or voltage generator is applied to appropriate pins and its voltage or current is measured. Results can be compared to predicted values of those of a known 'good board'.

An alternative method, using a tester with two roving arms, is to measure capacity. The board forms one plate of a capacitor by being pressed onto a dielectric layer over a plate. If there is a short circuit to other tracks, then the capacity is larger than it should be at a particular test point. If there is an open circuit, the capacity will be less (Conti 1986).

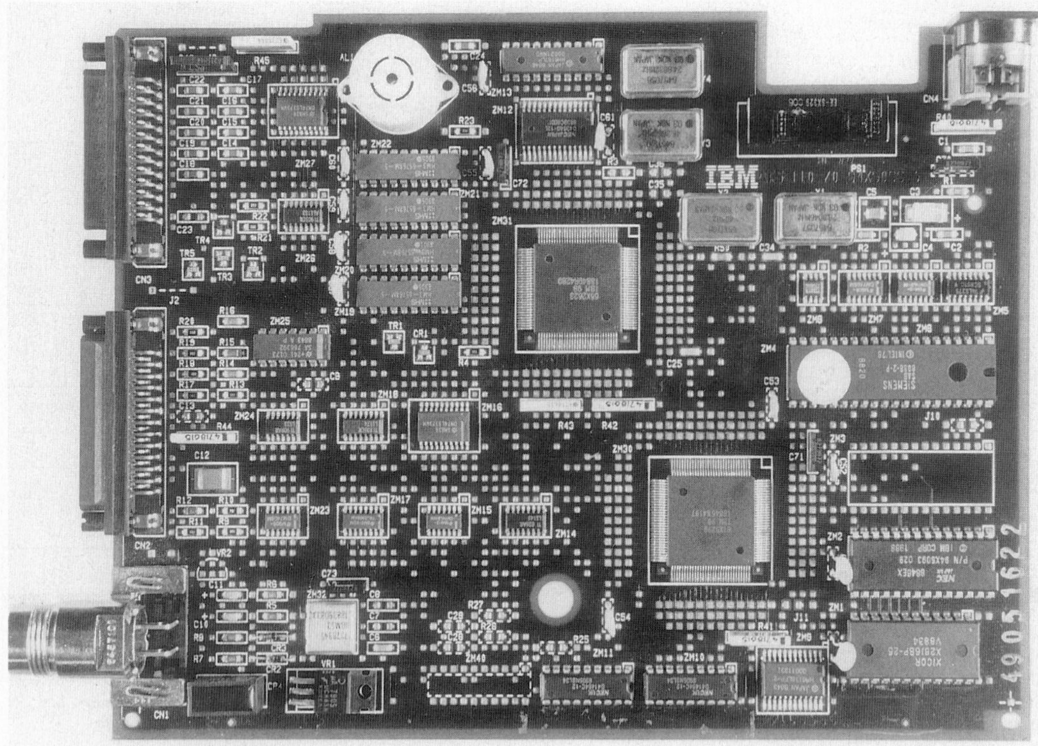

Figure 8.5 Test points on a surface mount printed circuit board

8.4.2 Incircuit testing

Incircuit testing checks the correct assembly of the board without undertaking a full functional test. There are three basic approaches (AEU 1986):

1. Impedance testing: the impedance between two test points is compared to those of a known good board.
2. Component testing: components are isolated by measuring operational amplifiers, applying a known voltage, and measuring the current magnitude and phase. This is illustrated in Figure 8.6. It makes use of the virtual earth property of an operational amplifier. This approach is particularly suitable for analog, as well as digital, circuits.
3. Digital integrated circuits testing: signals are applied and outputs monitored. Because the integrated circuit is not isolated and can be connected in feedback systems, the output for a device is frequently not what the manufacturer's data book states. This test can assist in ensuring the correct integrated circuit has been employed and that it has been connected in the correct polarity.

Thus the incircuit tester does not test the functionality of the board, but locates defects and incorrectly placed or soldered components. It has the advantage for rework in that the technician knows precisely which components need to be replaced.

8.4.3 Functional testing

While the functionality of boards can be tested by simply plugging them into the equipment in which they will finally be used, such methods do not fully test the card. For example, using a microprocessor card to run a particular system may not test all the instruction set.

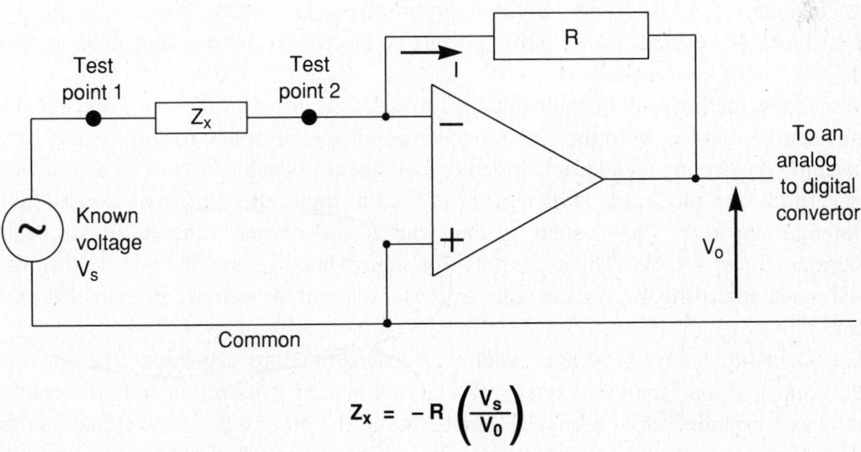

$$Z_x = -R\left(\frac{V_s}{V_o}\right)$$

Figure 8.6 Basic component tester circuit using a measuring operational amplifier

Functional testing allows the designer to generate a set of test vectors so that the full card can be tested (King 1987). The tests may be both static and dynamic. Alternatively, the input signals may be pseudorandom sequences and signature analysis may be used in a go/no go test. In all cases, either a special measuring system or a universal tester can be used. A growing difficulty with universal testers is the increasing speed at which they must operate, as the bed-of-nails contacts provide considerable inductance and stray capacitance.

The faults detected by functional testing are often similar to those isolated by incircuit testing. However, each method has its own distinctive range of fault types. Therefore, to locate as many faults as possible, both are used. The testing of boards should include all possible inspection and testing methods, as illustrated in Figure 8.7.

8.4.4 Built in tests

With the increasing complexity of printed circuit boards there is a tendency to use the technique developed for VLSI design—namely, designing for testability. The system has additional components included on each board so that the board/system can test itself. Called built in test (BIT) or built in self test (BIST), this technique minimizes the external hardware required for testing and enables the equivalent of incircuit and functional testing to be carried out on digital boards. There are three basic approaches that can be used (Haskard 1990):

1. bus system approach
2. scan method
3. signature analysis.

In the bus system approach, the board is organized as a bus structure with multiplexers that can switch each function block on or off the bus. Thus, it is possible to isolate function blocks and test them. Should a microprocessor be on the board, special software can be included in a test ROM for the microprocessor to perform these tests. As each function block is verified, it can either be left connected to the bus and used to assist subsequent testing or switched out.

In scan test methods, all bistable circuits are accessed using a serial shift register. This register can be formed by either interconnecting all the bistables in the system or by permanently connecting an additional identical number of bistable circuits as a serial shift register. In the test mode, the shift register is used to input all initial state conditions to the internal registers. The system is then run in its normal function mode, using predetermined test vectors. The system can be interrogated at any stage by returning to the test mode and using the shift register to reveal the current state of all bistables in the system.

The signature analysis system employs a pseudorandom sequence test generator (PRSG) and a signal analysis register (SAR), formed by connecting half the internal registers as a pseudorandom sequence generator and the other half as the signal analysis register. These registers, used to test the logic circuits on the board, operate on the principle that after a predetermined number of clock cycles, the state of the output should always be the same—that is, unless there is a fault in the system. This state is the signature for that particular logic function block. This system, in which the test determines if a particular digit pattern appears in the signal analysis register, can be organized as a simple go/no go test.

More recently, a proposal has been made by the Joint Test Action Group to produce a standard method of achieving BIT in VLSI packages and printed circuit boards (McLean & Romeu 1989).

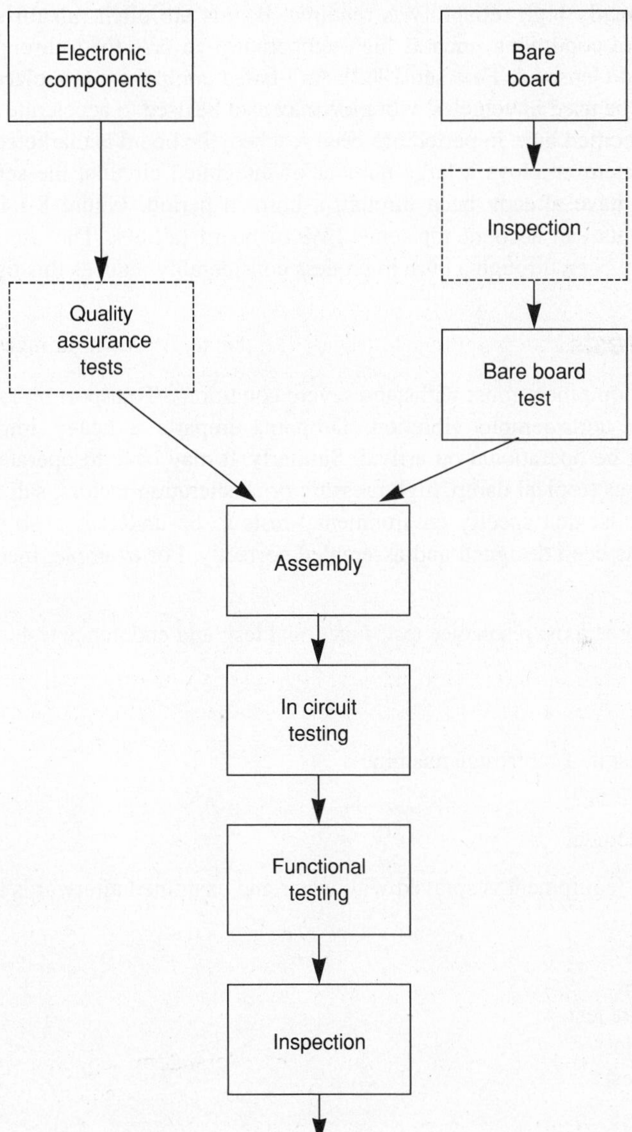

Figure 8.7 Flow diagram for the inspection and testing of a complex printed circuit board. Dotted boxes are optional steps

8.5 Testing after manufacture

To achieve and maintain the appropriate quality, two other types of testing are often undertaken: burn in and life tests.

8.5.1 Burn in

Where exceptionally high reliability is required, boards are often run for several hours under accelerated conditions, such as high temperature, to take them through the 'infant mortality' period (Jensen & Petersen 1982). Any failed component is replaced. When the equipment is to be used in vehicles, vibration may also be used to accelerate conditions.

Once the specified burn in period has been reached, the board is marketed.

Where a system employs a large number of integrated circuits, the semiconductors purchased may have already been through a burn in period. Figure 8.1 indicates that faulty components can account for some 13% of board failures. The use of integrated circuits that have been through a burn in process considerably reduces this figure.

8.5.2 Life tests

Many items of equipment must withstand severe conditions. Transport by rail, vehicle or air can impose considerable vibration. Dropping imparts a heavy impact, yet the equipment must be operational on arrival. Similarly, it may have to operate under harsh conditions such as tropical damp, high pressure or acceleration factors, salt spray, and so on. Standards exist that specify environmental tests to be undertaken on equipment to ensure that it has been designed and assembled correctly. For example, mechanical tests include:

- vibration on three axis: resonance test; functional test; and endurance test
- bump test
- drop test
- toppling test
- acceleration test in a centrifuge machine
- shock or impact test.

Climatic tests include:

- drip proof test (equipment is sprayed with water and examined afterwards for any water penetration)
- immersion test
- tropical life test
- low temperature test
- dust and sand test
- mold growth test
- corrosion test.

In addition to these tests to prototype products and first production runs (designed to show a customer that the product is of adequate quality), life tests are often carried out by organizations to establish their own quality procedures. Life tests may be made on assembled cards to ensure processing is correct. Usually thermal heat cycling with or without humidity (damp heat) is employed (Sinnadurai 1985; Sutherland & Videlo 1985), but more recently mechanical cycling methods have also been employed (Sinnadurai, Cooper & Woodhouse 1986). For example, much research has been carried out to establish correct design and soldering procedures for surface mount components. In the case of leadless ceramic chip carriers, the height of the solder pillars is critical to help decrease lift off during thermal cycling. Finally, such life tests are used to determine failure rates of boards and systems so that internal figures can be updated for use in the design of future systems.

8.6 Rework and repair

When boards are inspected and/or tested and defects are discovered, a decision must be made whether or not it is worth salvaging the board. In the case of very simple boards with few defects, it may not be economical to rework them. Today though, many boards are highly complex and a completed board is extremely valuable. It is therefore economically feasible to rework it so that it will pass all tests.

Further, boards that fail in the field may be returned for repair. In both cases, namely, rework and repair, the faulty component must be removed and replaced with a new one. This exercise is normally undertaken manually and, if done correctly, it should not be obvious that the component has been changed—that is, no board damage and all joints soldered correctly.

While the operation can be undertaken with the simplest of tools (a soldering iron and wicking braid to remove the solder), for surface mount components special rework tools are used. These may include heated tongs to remove passive components or hot air reflow soldering units.

When repairing boards, various pressure pack chemicals are often used. Care must be taken to use only those types that do not damage the environment. Mixtures are available for cleaning, moisture displacement, flux removal, wiper lubricants and freezing sprays to locate thermally sensitive components.

8.6.1 Component removal

In the case of wicking braid, a clean section of the braid is placed over the joint to be desoldered. The iron is cleaned, tinned and placed on the braid above the joint and a small heat bridge formed. The solder, on melting, is 'wicked' up the braid through the action of surface tension. Once all joints of the component have been desoldered, the component can be removed and the board cleaned. This method can be used with leaded and surface mount components. The braid comes in various widths and the appropriate width should be selected for the joints.

For leaded joints, a desoldering workstation can be employed. It consists of a temperature controlled iron with a concentric hole in the tip through which suction can be applied from a small vacuum system that is operated by a foot switch. The vacuum unit can also be reversed so that hot air is blown out of the tip to heat components, including heat shrink tubing. In the barrel of the iron or elsewhere in the machine, there is a container that collects the used solder. To use the desoldering workstation, the soldering iron is set to the normal temperature (approximately 310°C), cleaned and tinned. The hole in the center of the tip is also cleared. The iron is placed on the board joint with the hole in the tip over the end of the component lead. On melting the solder, the iron is wobbled in a circular motion and the vacuum applied with the foot switch. The solder is sucked into the machine, leaving a clean, solder free joint. Once all the component leads have been desoldered, the component is removed and the board cleaned.

Hand held solder suckers are a poor alternative. They consist of a plunger and spring release. When the solder is melted, the tip of the solder sucker is placed on the molten solder and the plunger released. The suction pulls the molten solder up into the chamber of the solder sucker. Unfortunately, the rush of air often cools the solder, solidifying it. To avoid this, the joint is often heated to an excessively high temperature. Figure 8.8 shows typical equipment used for manual desoldering.

Hot air rework stations are employed to remove surface mount packages (particularly, but not necessarily, high lead count integrated circuits). Figure 8.9 shows such a machine. It consists of a heating platform on which the board sits, possibly in an X-Y coordinatograph and, above it, a vacuum pick-up surrounded by a shroud that ducts the hot air onto the component leads. For each integrated circuit package type, the appropriate shroud must be used. Variables include the hot plate and hot air temperatures and the time the hot air is applied. An inspection microscope allows the operator to see when the solder has melted and use the vacuum pick-up to remove the component. The

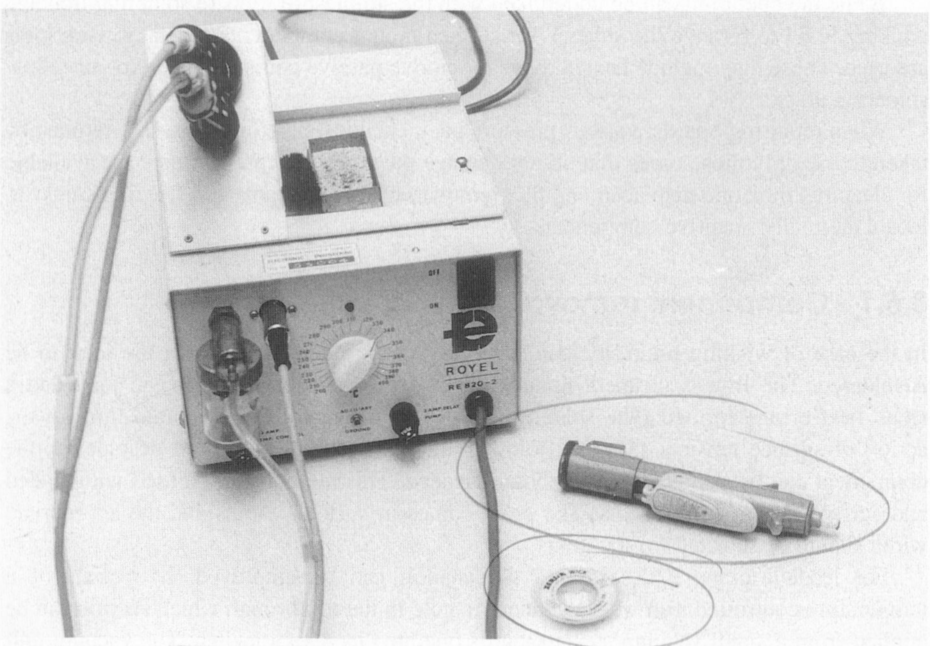

Figure 8.8 Equipment used to remove solder from a mounted component lead: (a) wicking braid; (b) solder sucker; and (c) desoldering workstation

board can be cleaned and, if need be, any excess solder removed with a hand soldering iron and solder wick.

Some boards with leaded or surface mount components may have a thick conformal coating over them to stop vibration. Before the component is replaced, this coating must be removed. This is done using a thermal parting tool, a hand piece that has a small resistance heated wire protruding from it. The heated wire is used to cut and peel away the coating around the component and the joint area. Once this is removed (often from both sides of a board for through-hole components), the component can be removed, as described above.

8.6.2 Component replacement

The mounting of the replacement component is undertaken manually, exactly as the placement and soldering of a component for the first time on the board. For hand soldering, see Sections 3.4 and 7.2.1. In the case of the hot air rework station, the reflow soldering methods discussed in Chapter 3 are employed. Solder paste is normally placed on the pad using a solder syringe dispensing unit. Using the vacuum pick-up and magnifier, the component is placed on the board in the correct position, the shroud lowered, and the hot air applied. The lower heating platform would normally be set to a temperature about 100°C.

8.6.3 Terminal removal

Often damaged turret and bifurcate terminals need to be removed. The obvious solution of drilling them out is not recommended because the board can be badly damaged when the drill catches and the pin rotates. The best method is first to remove any wires soldered to the terminal and then remove the solder fixing the pin to the board. Solder wicking braid is used for these jobs. The swaged end of the pin is now exposed and it is only

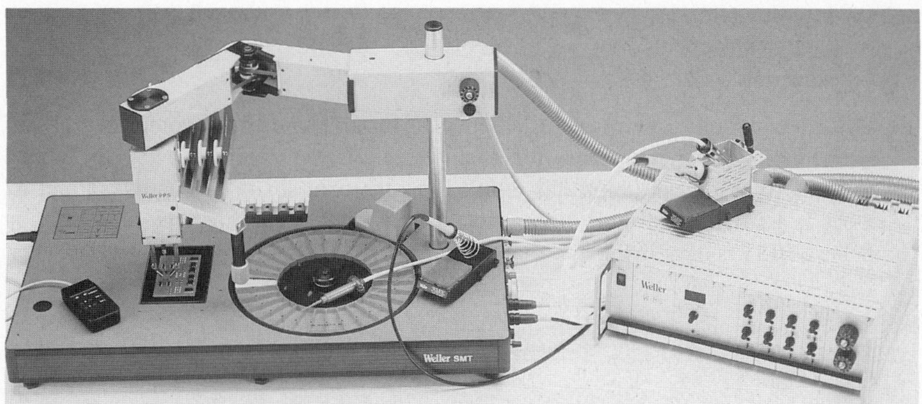

Figure 8.9 Hot air rework station
(Reproduced with permission of Cooper Tools Pty Ltd)

the mechanical friction that holds the pin in place. Using transverse end cutters, the swaged end of the terminal is cut away. The terminal can now be pulled out of the board, leaving an undamaged hole for the insertion of a replacement terminal.

The board is cleaned and a new terminal swaged and soldered in place, as described in Section 7.5.

8.7 Questions

1. Quality is an attitude; a total approach to business. What do you understand by this statement? Do you think it is correct?
2. Define the terms 'quality' and 'quality control'. Why is quality so important?
3. Through oxidization, the solderability of components decreases with age. What does this say about stock control procedures if a high level of quality is to be maintained? What does one do with 'old' components? (See also Section 9.5)
4. When inspecting a loaded board which uses mixed technology (both leaded and surface mount components), what defects would you look for?
5. Explain the differences and similarities between incircuit and functional testing.
6. Draw up a table comparing the bed-of-nails and roving arm methods of bare board testing. Which equipment type would you purchase for boards having:

   (a) 100 test points?
   (b) 1000 test points?
   (c) 10,000 test points?

7. You are to prepare a list of equipment needed for a board repair station. Divide your list into essential and optional equipment.
8. What are the objectives of life test and burn in processes? In what way do they differ?
9. Figure 8.1 classifies the different types of board faults. Have you any suggestions of how the following faults may be reduced?

   (a) insertion errors
   (b) faulty components
   (c) soldering faults

10. Even though built in self test (BIST) methods make boards more complex, their use is increasing in many areas; for example, electronic equipment for the armed services. Why do you think this is so? Does the complexity of boards warrant it? Does it assist in field fault diagnostics? Discuss these and other reasons why you believe BIST is of growing importance today.

8.8 References

Australian Standard AS3900-ISO 9000 (1987), 'Quality systems—Guide to selection and use'.

Australian Standard AS3901-ISO 9001 (1987), 'Quality systems for design/development, production, installation and servicing'.

Australian Standard AS3902-ISO 9002 (1987), 'Quality systems for production and installation'.

Australian Standard AS3903-ISO 9003 (1987), 'Quality systems for final inspection and test'.

Australian Standard AS3904-ISO 9004 (1987), 'Quality systems—Guide to quality management and quality systems elements'.

AEU (1986), 'Conditions for the selection of incircuit testers', *Journal of Asia Electronics Union*, **126** (July), pp. 20–4.

Beselin, K. & Schauflinger, H. (1987), 'Automatic testing of surface mount components', *Electronics and Wireless World*, **93**(1613), pp. 273–6.

Conti, J. A. (1986), 'Bare board testing', *Canadian Electronic Engineering*, **30** (November), pp. 44–8.

Haskard, M. R. (1990), *An Introduction to Application Specific Integrated Circuits*, Prentice Hall, New York.

Jensen, F, & Petersen, N. E. (1982), *Burn-in*, Wiley, Chichester, UK.

Juha, M. (1987), 'Automatic inspection: Are we solving the right problem—Parts 1 and 2', *Surface Mount Technology*, **1** (April), pp. 26–7 and (June), pp. 13–15.

Keely, C. A. (1989), 'Solder joint inspection using laser doppler vibrometry', *Hewlett-Packard Journal*, **40**(5), pp. 81–5.

King, H. (1987), 'Testing Customer-Specific Products', *Electronic Engineering*, **59** (June), pp. 37–44.

Lea, C., Howie, F. H. & Seah, M. P. (1985), 'Automated inspection of printed circuit board solder joints: An assessment of the capability of the Vanzette LI-6000 infra-red laser inspection instrument', *Brazing and Soldering* (8), Spring, pp. 34–42.

McLean, D. & Romeu, J. (1989), 'Design for testability with JTAG test methods', *Electronic Design*, **36**(12), pp. 67–71.

Pyzdek T. (1989), *What Every Engineer Should Know About Quality Control*, Marcel Dekker, New York.

Richards, B. P. & Footner, P. K. (1987), 'Is the incoming physical inspection of microelectronic components really necessary?', *IEC Journal of Research*, **5**(1), pp. 1–12.

Siemens (1986), 'Test strategies and procedures for SMD assemblies', Siemens AG, Munich, B9-133533-X-X-7600.

Sinnadurai, N. (1985), *Handbook of Microelectronics Packing and Interconnection Technologies*, Electrochemical Publications, Ayr, Scotland.

Sinnadurai, N., Cooper, K. & Woodhouse, J. (1986), 'Assessing the joints in surface mounted assemblies', *Microelectronics Journal*, **17**(2), pp. 21–31.

Sutherland, R. R. & Videlo I.D.E. (1985), 'Accelerated life testing of small-geometry printed circuit boards', *Printed Circuit Fabrication*, **8**(10), pp. 24–38.

Teraoka T. (1986), 'Incircuit tester saves time and labour in PC board assembly', *Journal of Asia Electronics Union*, **126** (July), pp. 26–32.

9 Production methods

9.1 Introduction to computer integrated manufacture

The methods employed to establish production lines are undergoing dramatic changes, mainly due to the introduction of computers and increased pressure to improve quality and reduce waste.

Computers have been used in the manufacture and design area for more than 20 years. Numerically controlled machines, combined with robotics, have allowed computer aided manufacture (CAM). The use of computers in product design, known as computer aided design (CAD), when combined with CAM, is often referred to as computer aided engineering (CAE). Computers are also used in many other areas, however, including finance and administration. Other computer support techniques are coming into play, such as optimized production technology (OPT) and manufacturing resource planning (MRP).

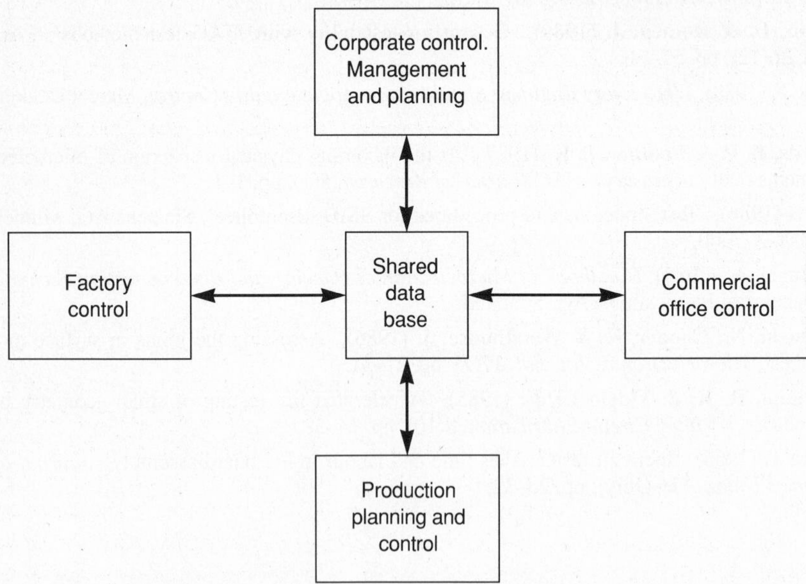

Figure 9.1 Communications and levels in a CIM system

The latter is a planning tool to handle flow of material and work, including inventory control and purchasing. The integration of all these systems is referred to as computer integrated manufacture (CIM).

Essential to any CIM system are an adequate network that allows communications (Patterson 1988) and a central data bank (Figure 9.1) containing (Sepehri 1987):

1. Product data: information on every product, be it mechanical and electrical specifications/drawings and pricing details.
2. Production data: details on how each product is manufactured.
3. Operational data: specific information on past, current and future productions, including lot sizes, scheduling information, etc.
4. Resource data: the resources needed, machines, people, and money for each operation.

Since this data is held on a shared data base, there must be sufficient security and checking built in to ensure that the data is accurate and reliable and only available on a need-to-know basis.

The design and establishment of a manufacturing system should include both top down and bottom up approaches and allow for a multilevel structure built up of modules. Passlow (1988) lists the steps:

1. an investigative study
2. establishment of a function specification and definition
3. a cost/benefit analysis
4. determining project management strategy.

Where possible, it should incorporate existing plant (Banister 1988). However, it must be stressed that it is essential to adopt design principles and development strategies that will satisfy needs over a reasonable lifetime, generally considered to be tens of years (Peters, Brackmann & Park 1987; Lau 1988). In the case of printed circuit board production, most of the equipment discussed in Chapter 7 can be integrated into such a system.

This chapter will concentrate on recent advances in manufacturing system design and several problems peculiar to the electronics industry.

9.2 Flexible manufacturing systems

All manufacturing systems should be flexible to ensure that they do not become outmoded. The term flexible manufacturing systems (FMS) refers to a specific process, however (Bonetto 1988). It involves the use of automation and robotic technologies to achieve a high level of flexibility (Peters et al. 1987). There are three principles for assembly lines: (1) planning, (2) scheduling and control, and (3) the continuous movement of products through the factory. Many organizations are required to manufacture either a large number of products in small volumes or a large volume product in small quantities at regular intervals. Rather than continually set up special lines for each different product type, it is better to establish a single line that is flexible enough to manufacture all the products. This can be organized on a small job lot basis (batch processing) or as a continuous flow. This is illustrated in Figure 9.2.

The advantages of FMS are:

1. The same or greater productivity often using less factory floor space.
2. Streamline flow of production using less operators, although more maintenance staff are required.
3. Continual 24 hour day operation with few penalties.
4. Consistency in product quality and reliability.
5. The ability to cope with a wide range of product types, allowing the company to move rapidly with new products.

These systems have four important disadvantages:

1. Increased capital cost to establish a new facility.
2. To operate reliably with zero breakdowns, preventive maintenance is essential. But this means that the time items of plant are down for maintenance can be higher than anticipated. Some component placement equipment requires weekly maintenance which can take up to several hours at a time.
3. Tooling costs can be high. For each new product, new solder screens, test head, handling carousels, or plates need to be produced. Further, the machines need to be programmed. This can be undertaken off-line, but with smaller FMS it is often done on-line which holds up production.
4. Finally, if the system is not designed so that it can be readily extended, it may have built in inflexibility so that it cannot accommodate future products. Examples include the setting up of a surface mount FMS that cannot cope with through-hole, leaded components or a batch FMS which is unable to cope with product volume growth. Perhaps a flow line FMS should have been established.

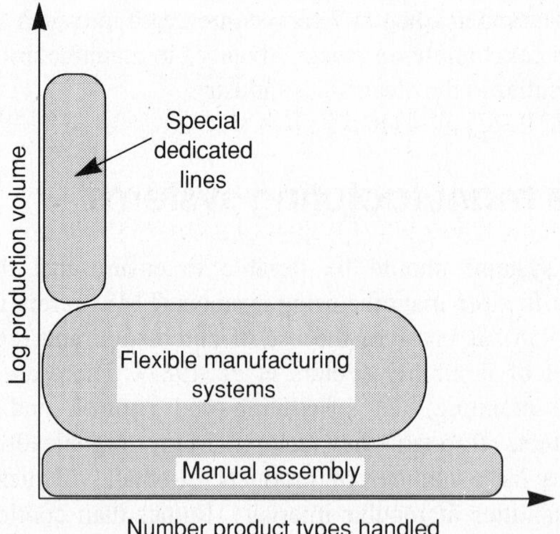

Figure 9.2 Assembly options available and their position for optimum production

Depending upon the range of products and volumes, there are various system options that can be implemented. If undertaken correctly, they allow the flow from one to the other as more automation is required:

- A flexible manufacturing module (FMM) is a simple machine that carries out more than one operation, often without human intervention.
- A flexible manufacturing cell (FMC) which consists of two or more modules, serviced by a common materials handling device or robot. It is usually fully automated and may incorporate features like automatic parts identification and inspection.
- A flexible manufacturing line (FML) is employed where there is a large similarity in products. This is often the case for the assembly of printed circuit boards. The line consists of many machines in sequence, serviced by automatic material transport and handling devices. There is a continuous flow of materials from input to output so that the line can handle large quantities of a range of similar products.
- A batch flexible manufacturing system (BFMS) is similar to FML, but organised on a batch basis so that it can handle a large number of different products in small batch quantities. There is a computer controlled transport or robotic system to interconnect a number of flexible and versatile machines. Figure 9.3 shows such a system operating at British Aerospace for surface mount boards. It consists of three cells; the first carries out the pre-clean, solder print, and component placement. In the second, the components are reflow soldered in a vapor phase reflow solder unit and the boards cleaned to remove flux residue. In the third and final cell, the boards are tested. Boards are transported in carousels in batch quantities of seven. Robotic arms and an overhead transport system move boards and carousels respectively. An automatic component storage unit is also housed in the same area.

During manufacture, it is necessary to identify each major module, such as a printed circuit board. It is usual to use a bar code or similar code so that machines can automatically identify each board as it enters a cell or process. Cassettes/carousels are also coded to identify batches (McClintock & Nosier 1985).

9.3 Establishing a manufacturing system

Production lines and systems are normally established so that there is a clean flow of raw material in and finished product out. The three golden rules are:

1. zero breakdown of plant
2. zero product reject levels
3. zero stores, so that all components are being assembled and finished products marketed.

At the design stage engineers examine a product, break it down into its basic parts and then lay out the factory floor area into a compact manufacturing and assembly line using, whenever possible, existing equipment and people. Today the planning and establishment of production facilities, be it a FMS or dedicated line, can be undertaken through simulation programs such as SIMFACTORY. Having itemized the production steps for a product or range of products and accounted for the specifications of plant equipment, the

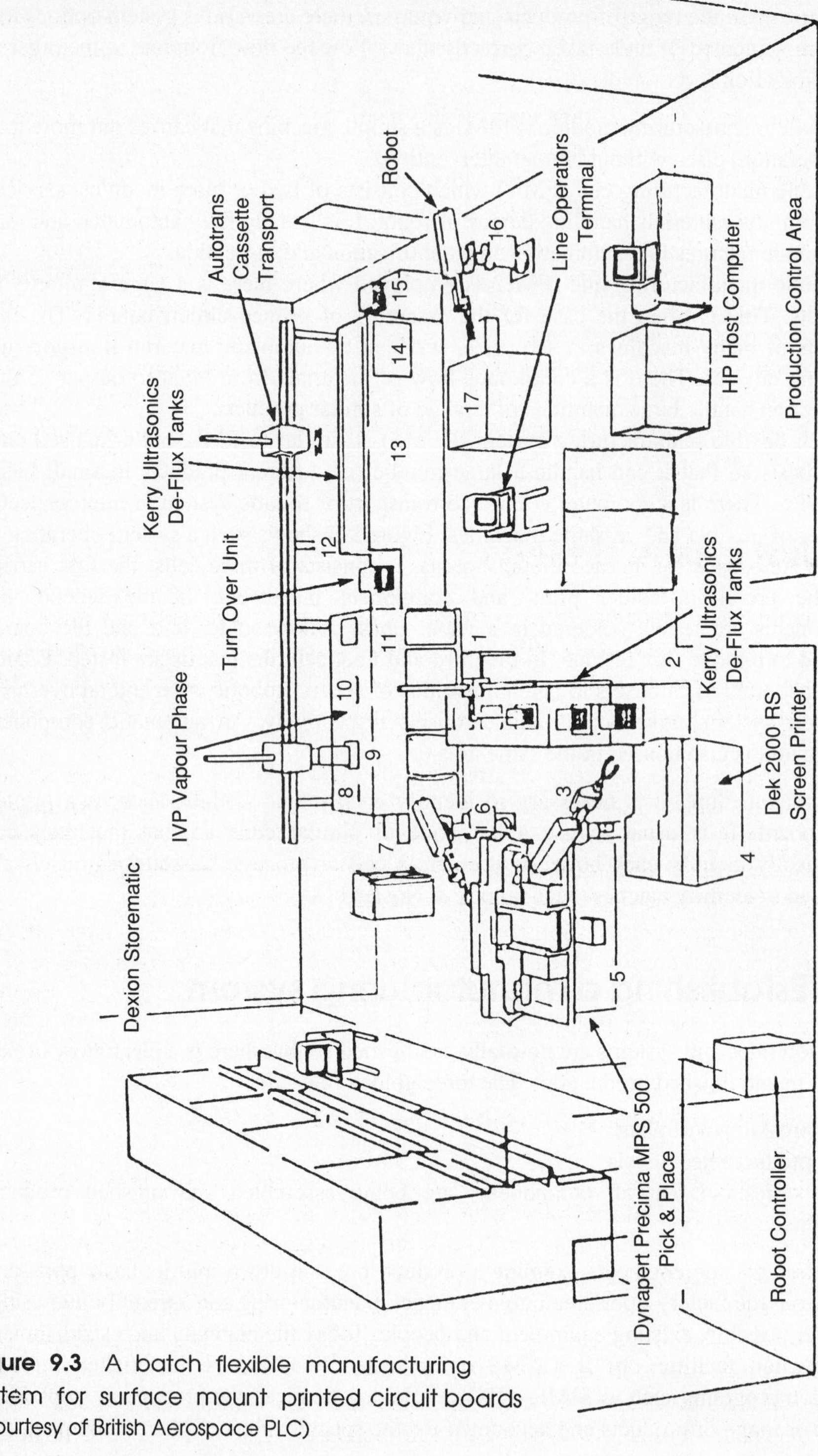

Figure 9.3 A batch flexible manufacturing system for surface mount printed circuit boards (Courtesy of British Aerospace PLC)

software models production runs. Different equipment types and quantities can be used and assessed in terms of efficiency, bottlenecks, etc. Comparisons can be made between manual, partly automated and full FMS for different product types and volume levels. The program uses icons so that the operator has on display a visual picture of the factory layout and operation. After a specified run time, detailed information can be extracted in tabular form. The system allows a complete breakdown on the performance of any machine, module, robot or transporter, storage requirements, etc. Factors such as the number of personnel required, impact of random breakdown of plant equipment, transporter router strategies and length of queues can be examined. These simulation programs are therefore extremely powerful tools for designing and verifying the operation of a manufacturing system.

At present there is one serious difficulty in that not all manufacturers of plant are prepared to provide all the relevant details that the software requires. As factories make more and more use of these programs, however, pressure on plant manufacturers will force them to supply the necessary information.

9.4 Just in time strategy

The just in time (JIT) approach aims to control all resources and eliminate all waste in a manufacturing system (Mortimer 1986; Hay 1988). Waste is defined as anything that is not necessary to accomplish the object in hand, including having buffer stock, excess capacity, and additional people. Only the required amount is manufactured and only when it is needed.

It also encourages a consolidated production system in which all machines are so grouped that all unnecessary transportation is removed. To prevent loss of production time from plant equipment failure, total preventive maintenance is advised. Employees are cross-trained to increase the flexibility of the system. There is also a full production control software tool to assist those planning and controlling manufacture.

Conceptually, the strategy has much to commend it, but care must be taken where there is a dependence upon others to achieve some objective. Because products are produced only as required, any hold up can make the process critical. Supply of external component parts is one area that must be watched closely.

9.5 Component storage

Storage areas are dead space and storage stocks tie up valuable capital. They, therefore, should be minimized. Unfortunately, with electronic components supplied on tapes or reels, the minimum quantities that can be ordered necessitate that some components be held in store. Thus storage cannot be eliminated. Inventory systems for store are available which include purchasing and sales aspects. Adequate lead times must always be allowed and, in spite of the JIT strategy, it is often realistic to hold in store a supply of components which can be difficult to obtain.

As mentioned in Chapter 8, as components age, the solder oxidizes and their 'wettability' deteriorates so that even after a few months storage there can be difficulties in soldering them to the board. Some components are delivered in nitrogen sealed bags to eliminate this problem. There are, however, special inert gas storage cabinets available to MIL standards, to ensure that materials, components, and work in progress do not oxidize. The cabinets also hold the components in a low humidity environment and offer various levels of security and electrostatic protection.

9.6 A clean assembly environment

To achieve quality products, cleanliness is essential. Not only the components, boards, soldering process and equipment, but the environment must also be clean. The level of cleanliness of the assembly/manufacturing environment depends upon the product

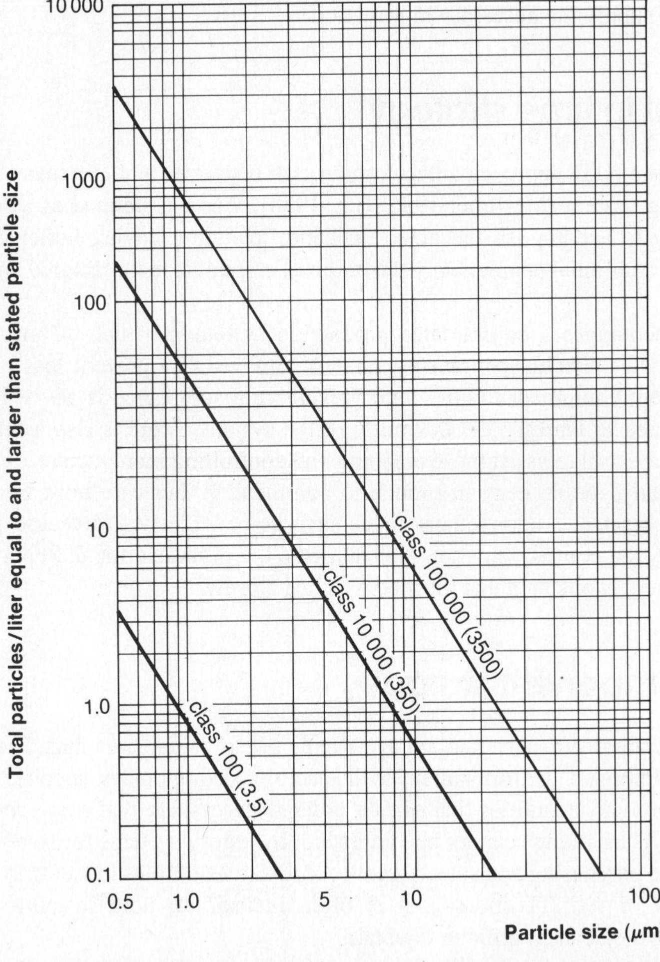

Figure 9.4 Particle size distribution for three clean room standard classifications

market—namely, the quality and reliability a customer is prepared to pay for. With consumer products, normal air conditioning standards are adequate. For some professional and armed services work, a clean room is necessary while space and medical electronics work demands the ultimate in clean rooms. Various clean room standards documents are available in three or more classifications: 100,000 (3500), 10,000 (350) and 100 (3.5). The figures indicate the number of particles of a half micron size or larger per cubic foot (per liter). Temperature and humidity limits must also be controlled.

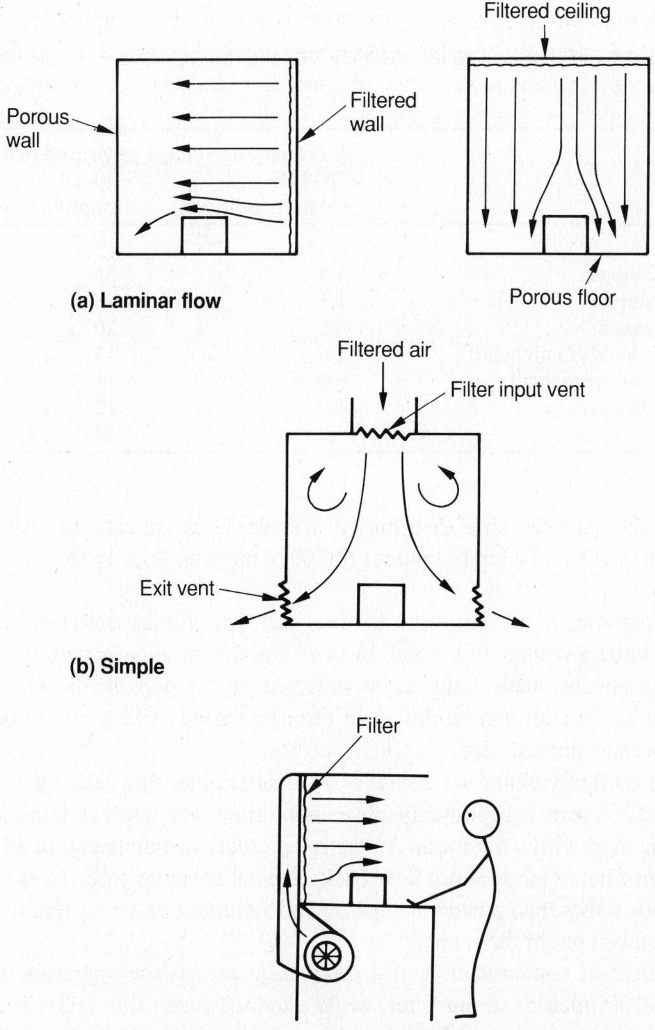

Figure 9.5 Clean room methods: (a) laminar flow, (b) low cost version and (c) laminar flow work bench

Table 9.1 Quantities of particles shed by humans

Activity	Typical quantity of particles shed by an adult per minute
Still, no movement	10^5
Sitting, slight head and hand movement	5×10^5
Sitting, with increased movement	10^6
Rising from sitting to standing	2.5×10^6
Average walking speed	7.5×10^6

Table 9.2 Electrostatic voltages generated by various operations for two levels of relative humidity

Source	Electrostatic voltage generated in kV 70–90% relative humidity	10–20% relative humidty
Walking on vinyl floor	0.25	12
Walking on synthetic carpet	1.5	35
Sitting on a foam cushion	1.5	18
Picking up standard plastic bag	0.6	20
Pulling tape from a printed circuit board	1.5	12
Skin packing a printed circuit board	3.0	16
Cleaning a circuit with an eraser	1.0	12
Freon circuit spray	5.0	15

Figure 9.4 shows the particle size distribution for three classifications (Australian Standards Document AS1387, Federal Standard FS209, American Society for Testing and Materials, F318).

The number of particles in a clean room is measured with a light scattering monitor. The unit draws in a known volume of air and, in so doing, the air with its particles passes through a beam of monochromatic light. Light reflected by the particles is sensed by a photo cell and counted. The air can be drawn in through various size filters to obtain a histogram of count versus particle size.

Clean rooms are normally either horizontal or vertical laminar flow (Austin 1970), as shown in Figure 9.5. Where exceptionally clean conditions are needed, laminar flow work benches can be used within the room. A positive pressure is maintained in all rooms to ensure outside non-filtered air does not flow back. Special cleaning procedures must be adopted, using liquids rather than powders and a ducted vacuum cleaner system to ensure that all particles are taken out of the room.

The greatest source of contamination of a production area is the operators. Human hair is typically 30–100 microns in diameter, while dandruff (dead skin cells) and other particles shed by people range from 0.5–500 microns in size. Table 9.1 indicates typical quantities of these particles shed every minute. To contain them, some protective garment should be worn, often with head, face, and feet covering. Special lint-free material is used to make the garments (Australian Standards Document AS2013.4, American Society for Testing and Materials F51).

Table 9.3 Sensitivity of various component types to ESD voltages

Device type	Range ESD sensitivity (kV)
VMOS	.03–1.8
MOSFET	0.10–0.2
GaAs FET	0.10–0.3
EPROM	0.10–2.5
JFET	0.14–7.0
OP AMP	0.19–2.5
CMOS	0.25–3.0
TTL	0.30–2.5
Film resistors (thick, thin)	0.30–3.0
Bipolar transistors	0.38–7.0
ECL	0.50–1.5
SCR	0.68–1.0
Schotty TTL	1.00–2.5

9.7 Electrostatic protection

Many modern electronic components are damaged by normal electrostatic discharge (ESD), from both objects and humans. Table 9.2 presents typical voltages that may be generated (Fuqua 1983) while Table 9.3 gives typical voltages that cause components to fail (Gregg 1989).

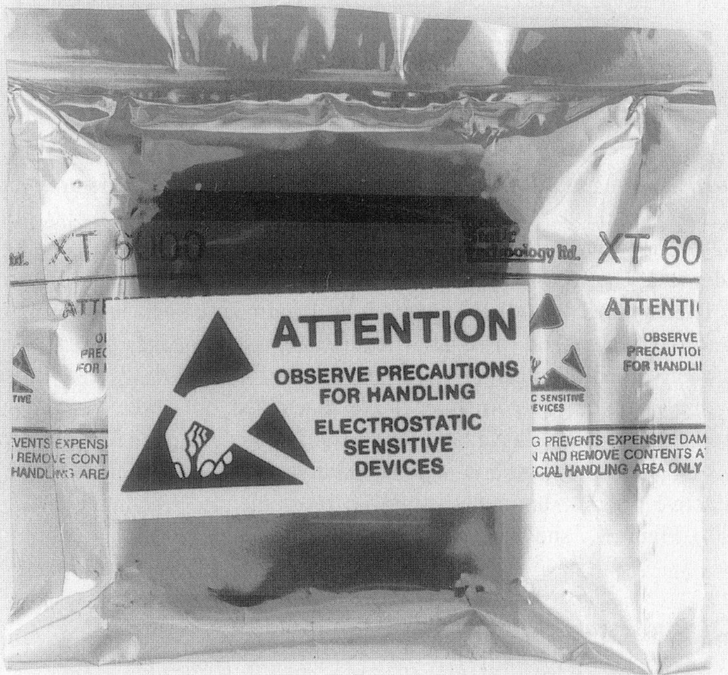

Figure 9.6 Antistatic bags for shipping semiconductor components

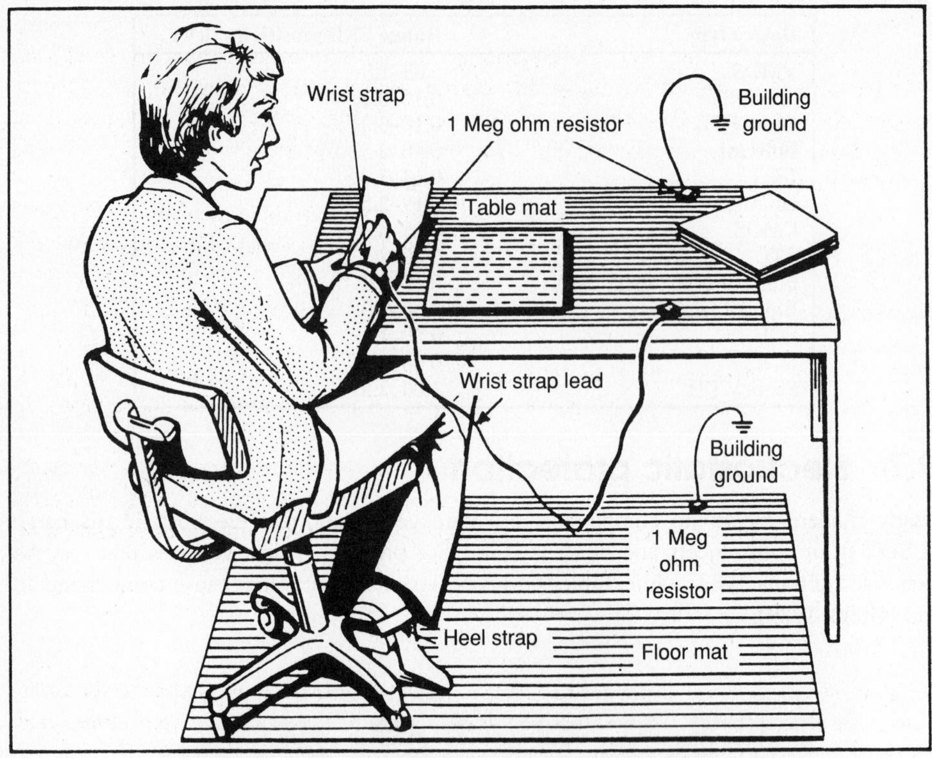

Figure 9.7 A typical antistatic work station

Two tests are often used to determine the sensitivity of semiconductor components to ESD. The first, conducted according to MIL standards, involves a 100 pF capacitor discharged through a 1500 ohm resistor (equivalent circuit of a human being) into the device under test. The second, of Japanese origin, uses a 200 pF capacitor and omits the resistor. It is equivalent to a discharge from an inanimate object. In spite of the inbuilt protection to the circuits and these test methods, the high voltages generated under low relative humidity conditions can damage components. Consequently, care must be taken to ensure that no damaged components are assembled on cards. Protective methods include special metal lined plastic bags (Figure 9.6), antistatic mats for bench tops, leg or wrist straps to earth operators, and antistatic sprays. In the construction of the assembly area, conductive floors, mats and seating are recommended (Antinone 1987; Kerns & Riskin 1984). Figure 9.7 shows an antistatic work station.

9.8 Questions

1. Computer integrated manufacturing (CIM) systems must have a good computer communications network. Why?

2. What is a flexible manufacturing system (FMS)? Why do factories make use of these systems?
3. A difficulty in setting up a manufacturing system is to make it flexible enough to cope with new products, extensions to current assembly techniques and variations in product quantities. Consider the implications of this problem when setting up a flexible manufacturing system. What solutions can you propose?
4. What do you understand by the term just in time?
5. Pick-and-place machines hold large quantities of components in reels and stick tubes. In effect, they become the component store. What impact does this have on component procurement and the philosophy outlined in Section 9.3 of zero stores?
6. What special precautions must be taken when setting up either a printed circuit board assembly, rework or repair area for military work?
7. Refer to Figure 9.3. Follow the sequence around from the starting point at Station 1 to the final test at Station 17. Make a sequential list of the processing steps for this line.
8. An important aspect of quality is cleanliness—environment, processes and component parts. Systems constructed for medical and space applications need a high degree of cleanliness. What steps would you take to ensure sufficient cleanliness of components, processes and environment for lines that are to assemble:

   (a) intensive care monitoring systems?
   (b) heart pace makers?

9. Some GaAs MESFETs are damaged if the gate drain/source voltage exceeds a few volts. What precautions need to be taken to ensure that they are both stored and assembled onto boards without damage?
10. One problem with microwave devices is that they often require the use of a more expensive low melting point solder for assembly. How would you adapt your reflow surface mount assembly line to accommodate this requirement, remembering that there are other normal components to be assembled onto the same board?

9.9 References

Antinone, R. J. (1987), 'How to prevent circuit zapping', *IEEE Spectrum*, **24** (April), pp. 34–8.

Austin, P. (1970), *Design and Operation of Clean Rooms*, Business News, Detroit.

Banister, R. (1988), 'New machines for old', *IEE Review*, **37**(7), pp. 283–7.

Bonetto, R. (1988), *Flexible Manufacturing Systems in Practice*, North Oxford Academic English Translation, London.

Fuqua, N. B. (1983), 'Static zap makes scrap', *Hewlett Packard Bench Briefs*, March–May, pp. 1–7.

Gregg, P. (1989), 'A static shock for electronics results in loss of money', *Electronics Weekly*, (1447), p. 12.

Hay, E. J. (1988), *The Just-in-time Breakthrough*, John Wiley, New York.

Kerns, R. C. & Riskin, J. R. (1984), *ESD Prevention Manual*, Analog Devices, Norwood, Mass.

Lau, G. (1988), 'CIM needs a long term strategy', *What's New in Processing Engineering*, **2**(1), pp. 32–3.

McClintock, J. C. & Nosier, R. W. (1985), 'Robots and data: Keys to quality electronic assembly', *ASQC Quality Congress Transaction*, Baltimore, pp. 100–13.

Mortimer, J. (ed.) (1986), *Just-in-Time: An Executive Briefing*', IFS (Publications), Bedford, UK.

Passlow, D. (1988), 'Successful implementation of CIM', *What's New in Processing Engineering*, **24**(1), pp. 26–8.

Patterson, G. (1988), 'Integrated PCB production', *OEM Design*, December, pp. 25–6.

Peters, L. S., Brackmann, E. J. & Park, W. T. (1987), 'Future directions in automation and robotics for manufacturing', *IEEE AES Magazine*, **2** (February), pp. 12–16.

Sepehri, M. (1987), 'Integrated data base for computer integrated manufacture', *IEEE Circuits and Devices*, **3** (March), pp. 48–54.

10 Design of a circuit card

10.1 Introduction

The generation of a new product employing electronics normally requires the design and production of one or more printed circuit cards. The product idea results from a marketing division's assessment of a new product, the need to upgrade an existing product, or negotiations with an external customer for a product. After the product specification and costing are finalised, the electronic portion normally goes through a design and development phase which results in a schematic diagram. Depending upon product volumes, there may or may not be a pilot production run, but eventually there is a satisfactory circuit and the need to design a final production card. In order to interface into the plant to manufacture and assemble these cards, a considerable amount of information must be supplied. Further, to do this correctly, an approved method of documentation and change control procedures must be established. Figure 10.1 shows in flow diagram form the final steps in the design, once the schematic diagram is known.

As Figure 10.1 shows, the design of the card is considerably more than a layout. The steps are:

1. Layout: involves the decisions on board style, through-hole, surface mount, or mixed technology; single, double, or multilayer boards; placement of components, vias, lands, test pads and device foot prints; and generation of the interconnection tracks, etc.
2. Design: undertaken in parallel with the layout to ensure that current densities in tracks are not excessive, track separation is adequate so there is no electrical breakdown, no hot spots on the board (thermal analysis), and that stray parasitics, principally stray capacities, do not prevent the circuit from operating correctly (extraction and simulation).
3. Net list: a software list that gives all interconnection paths. It can be compared with a net list generated by the schematic diagram to ensure no interconnection has been omitted or is incorrect.
4. Card artwork: a test plot of the final card layout which is used in the manual inspection of any components and the visual inspection of completed cards (Figure B.4 for example). Further, the via information and the net list is used to generate bare board testing information.
5. Photo tooling: software to drive the photo plotter to generate the appropriate photo negative or positive.

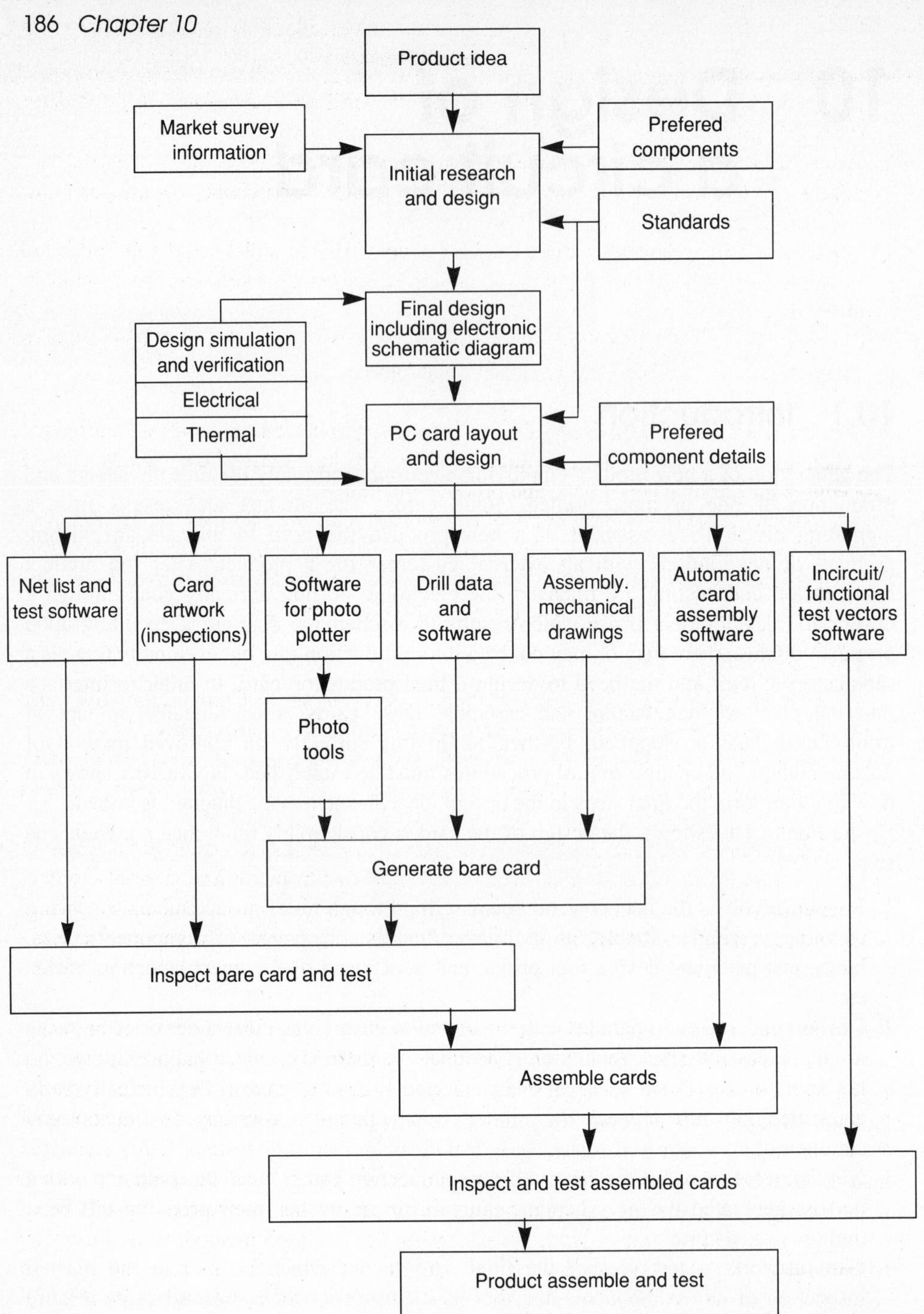

Figure 10.1 Steps in the design and production of a board showing the wide ranging information required from a designer to manufacture and assemble a board

6. Drill data and software: cards must be drilled not only with the correct drill size, but the resulting hole must be in the correct position. Software is therefore needed to drive the NC drilling machine and specify the drill sizes to be used.
7. Assembly drawing: the mechanical design information of the card. It includes card material, thickness, copper weight, overall dimensions and shape, datum positions, component size and position.
8. Automatic card assembly software: software which drives radial, axial and integrated circuit insertion machines or pick-and-place machines so cards can be assembled correctly.
9. Incircuit/function test software: allows the final testing and verification of the board. The software also includes the test vector generation.

At the conclusion of the card design phase, software and documentation must be provided for each of these nine steps. The documentation method must allow change control procedures for possible corrections and product updates.

An important aspect of this exercise is standards. Not only do they define preferred methods and give practical information on how to calculate design figures, but they also provide a wealth of background information, including details on a range of options. Overall, standards are concerned with ensuring quality of a product.

Standards apply at three levels:

1. International level, such as the International Electrotechnical Commission (IEC) and the Institute for Interconnecting and Packaging Electronic Circuits (IPC).
2. National level, including country standards such as American Standards (IPC, MIL, etc.), Australian Standards (AS), block community standards such as the European Community's Cenelec Electronic Components Committee (CECC).
3. Company level, which may simply be a document referring to standards in either of the above two categories, a detailed standard in its own right, or a mixture of the two. Government bodies, such as telecommunications and defense, and large private organizations tend to have internal standard documents that are based on international and national standards.

Standard documents take various approaches. Some are rigid in that they define precisely how something must be done. There are no options. This category includes those that register electronic device package styles such as the Joint Electron Device Engineering Council (JEDEC). Many company standards follow this pattern to ensure products are of the correct quality and will interface with their manufacturing system. Other standards aim to establish the quality level of a manufacturer rather than the product; once a manufacturer has achieved a certain quality certification then their products will be of that approved quality. This approach is becoming the accepted method in the European Community (CECC). Still yet another approach is a standard that recommends how to manufacture a product. This is the approach taken by IPC. Such documents are a useful source of information because they provide a range of alternative methods and details. The final standards approach is to establish a series of tests that identify poor quality products. This is the approach taken by the armed services such as the United States' Armed Services (MIL) documents.

The increasing demands placed on printed circuit board designs by increased packing densities, range of device types, standards and information needed to establish an automatic or semiautomatic production line, have meant that computer assisted layout is becoming the norm. In this chapter the underlying principles of card design and computer aided design (CAD) methods will be considered.

10.2 Layout of a card

The layout of a card involves the placement of components on the card and routing interconnection wires in such a way that the board is pleasing to the eye, compact and has the minimum number of interconnecting layers. To achieve this three main areas must be examined:

1. electrical
2. thermal
3. mechanical.

While all three are interrelated, they will be considered separately. First, a word on board datum systems.

All holes and pads in the board must be aligned according to a datum system. This is particularly critical where several layers are to be used because in manufacture every layer must align with all others. A suggested datum system is shown in Figure 10.2.

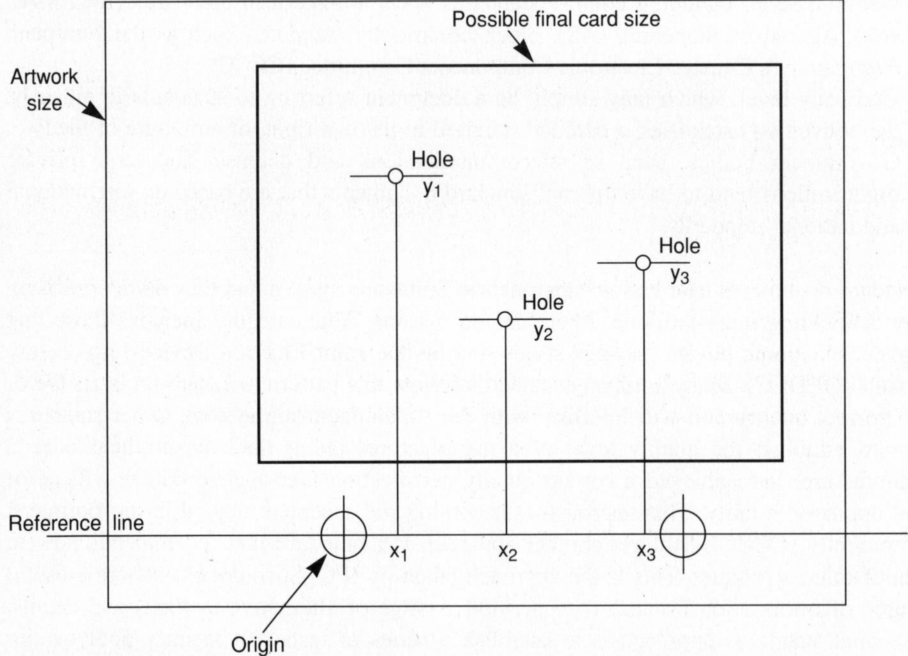

Figure 10.2 Datum and reference line for a printed circuit board

It uses two alignment patterns and a datum or a reference line. Special alignment/datum marks are used. All the component layout is based on these marks. More detailed considerations can be found in standards documents such as Australian Standards AS2546.3.

10.2.1 Electrical considerations

The requirements for a layout include:

- Minimum stray coupling and stray components, including minimum lead resistance, inductance, and shunt capacitance.
- Sufficient track separation to ensure that breakdown does not occur.
- Tracks of adequate width to carry the required currents.

The simplest way of starting a layout is to place components on the board as they are placed in a well laid out circuit schematic diagram. Normally this diagram has been arranged to minimize the length of interconnecting wires so that it appears open and clean. If a bus system is used, care must be taken to ensure all bus wires can be taken off the board via a plug/socket and that all blocks on the board can interface into it. Some computer aided design systems employ a rubber banding method so that the interconnecting wires are shown point to point and components can be moved around to minimize wiring complexity.

The resistance of any interconnection consists of two portions—the copper track and any vias or plated through-holes. The bulk resistance (ρ) of copper is typically 1.8×10^{-6} ohm/cm at 25°C, increasing with increasing temperature. The track resistance R_T is given by:

$$R_T = \frac{\rho L}{t W} \qquad [10.1]$$

where L is the track length
W is the track width
t is the track thickness, which depends upon the weight of the copper used and can be 18, 35, 70 or 105 microns

Thus at 25°C, the length of 4 inch (100 mm) of track of 1 oz copper (35 microns thick) and 40 mil (1 mm) wide is 50 milliohm.

For calculations at 100°C, the resistance increases by a factor of 1.25.

The resistance of plated through-holes depends upon the hole diameter and plating thickness. This is illustrated in Figure 10.3. The resistance of the hole, R_p, is:

$$R_P = \frac{\rho T}{\pi (d - t) t} \qquad [10.2]$$

where ρ is the bulk resistance of the copper and equals 1.8×10^{-6} ohm/cm at 25°C
D is the hole diameter (drilled)
t is the thickness of the plating
T is the board thickness

For a 40 mil (1 mm) hole in a 1⁄16 inch (1.6 mm) thick board, which is plated to 30 microns thick, the resistance is 0.3 milliohm at 25°C.

An alternative approach, sometimes used to reduce errors resulting from track and plated hole resistance voltage drops, is to employ Kelvin connections. Extra contacts are added to separate the measuring circuit from the larger circuit currents and associated voltage drop (due to track resistance), thereby eliminating their effects.

In some instances, it is necessary to limit the effects of very high short circuit currents or surge currents. The overload strains the bonding of the copper to the board because mechanical forces are produced due to the coefficient of thermal expansion of the copper. Should surge currents produce energy levels of 60 A.millsec or greater, for 1 oz copper (35 microns thick), track widths greater than 40 mil (1 mm) must be employed or standards documents, such as Australian Standard AS2546.3, should be consulted.

The voltage permissible between conducting tracks on the surface of a board before breakdown occurs depends on many factors such as board material and surface finish, coating (if any), environmental conditions, and track spacing. Most modern semiconductor circuits operate at voltages (≤ 24 volts) where this is not a problem. If higher voltages are used, spacing can become critical. For a normal uncoated epoxy card, operating at an altitude less than 3300 feet (1000 meters), spacing should be 0.8 mil/volt (0.002 mm/volt). This figure can be increased further for harsher conditions to ensure adequate safety. On some boards, mains supply voltages are used and the equipment must meet all relevant standards or supply authorities will not approve the item of equipment.

Shielding is often an important consideration in laying out a board. It may be needed to provide isolation, a ground plane for radio frequency circuits or a line of known characteristic impedance to prevent reflections (see Chapter 11). Where isolation is required, earth lines may be run on the board separating the lines that need to be isolated. Guard rings may be placed around the input to high gain analog circuits. Where large earth planes are used, the thermal mass of the plane can prevent satisfactory soldering to a pad, so a modified pad is often used. Further, large ground plane areas are often made by using crosshatching. Examples of shielding are given in Figure 10.4.

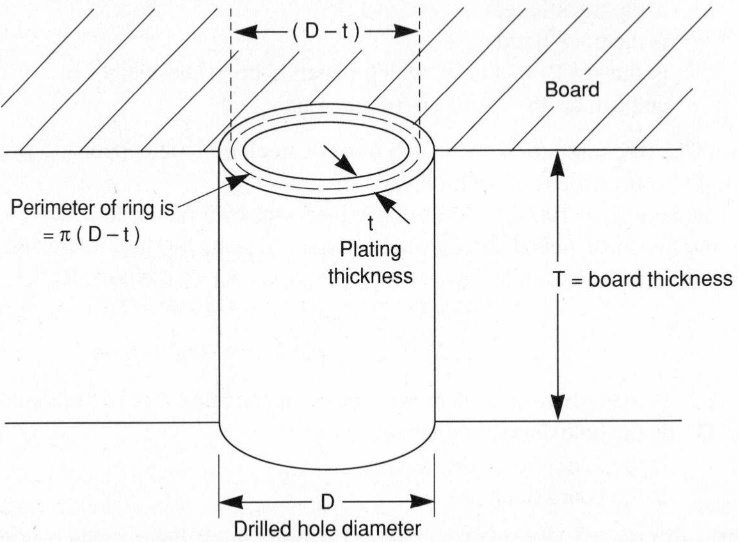

Figure 10.3 Plated through-hole dimensions

With multilayer boards, some conductor and heat sink layers can provide shielding and grounding. Thus signal planes can be interspersed with ground and supply lines to provide effective shielding.

10.2.2 Thermal considerations

There are two sources of heat: the board itself and the components mounted on it. Since each board type (and component) has a maximum temperature of operation, care must be taken to ensure that this temperature is not exceeded. Heat is removed by radiation, conduction and/or convection, depending upon the board construction and system thermal design. For example, boards may have heat pipes embedded in them or forced air cooling may be used in the cabinet in which the boards are to be mounted. Power devices

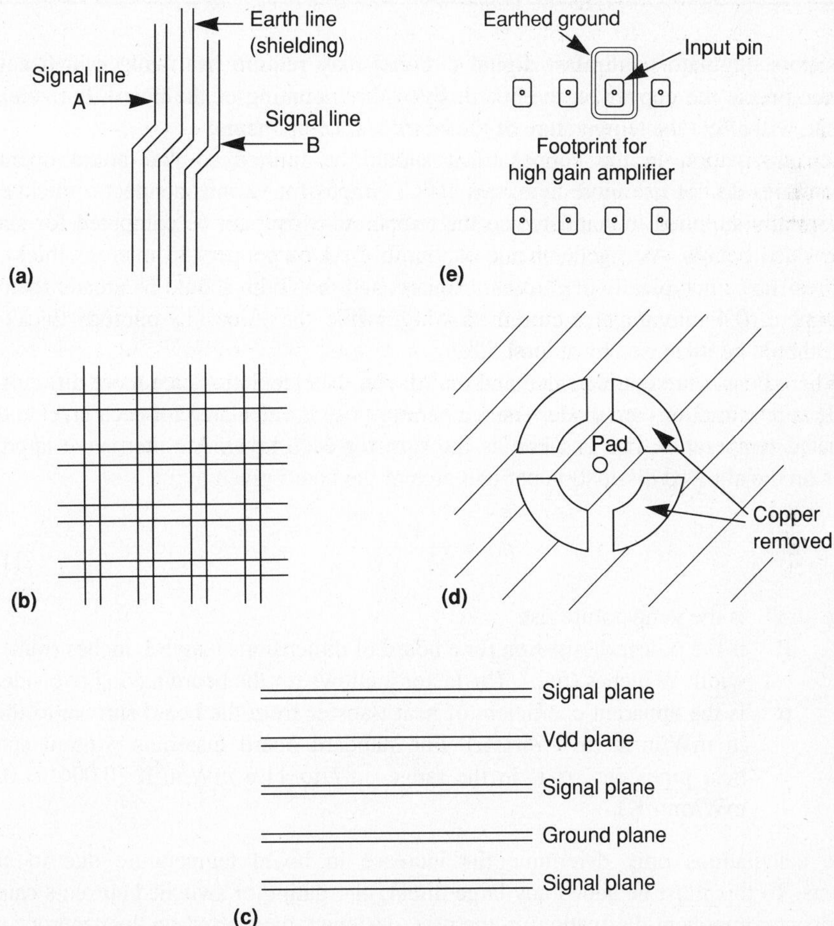

Figure 10.4 Examples of shielding and grounding: (a) shielding of two signal lines, (b) crosshatched earth plane, (c) use of layers in a multilayer board to give shielding, (d) earth plane pad shape, and (e) a guard ring for the input to an amplifier

Table 10.1 Minimum copper track width (mm) to ensure temperature rises less than 10, 20 and 40°C for various DC currents. M indicates dimension is small. Use minimum allowable track width

DC current amps	Temperature rise °C					
	½ oz (18 micron thick) copper			1 oz (35 micron thick) copper		
	10°C	20°C	40°C	10°C	20°C	40°C
0.1	0.03	M	M	M	M	M
0.5	0.25	0.3	0.1	0.15	0.1	0.06
1.0	0.625	0.5	0.3	0.4	0.25	0.15
2.0	1.6	1.0	0.75	0.8	0.5	0.3
5.0	7.0	4.0	2.5	3.25	1.75	1.0
10.0	—	10.0	5.6	8.0	4.7	3.0

(transistors, regulators, ultrafast digital circuits) may require heat sinks which can be separate or use the copper on the board. Even the mounting of the board, horizontal or vertical, will affect the temperature of the board and components.

The dissipation in the copper track should be limited so that board operating temperatures do not rise more than, say, 10°C. Graphs for various conductor thicknesses are normally supplied in standards so the temperature rise can be computed for simple single-sided boards. As a general rule of thumb for 1 oz copper (35 microns thick), the sheet resistance is typically 0.55 m ohm/square, and the width should be greater than 160 mil/ampere (0.4 mm/ampere) current flowing, while for ½ oz (18 microns thick) this width should be increased by at least 50%.

Where boards are double-sided and multilayer, the calculations are more difficult and simple approximations are made. The temperature rise is calculated for each layer and the estimated temperature rise is taken as the sum for each layer. An alternative approach works on the allowed dissipation per unit area of the board given by:

$$\Delta T = \frac{P}{2\,L\,W\,\alpha} \qquad [10.3]$$

where ΔT is the temperature rise

P is the power dissipation for a board of dimensions length L inches (mm) and width W inches (mm). The factor 2 allows for the board having two sides.

α is the apparent coefficient of heat transfer from the board surface to the air in mW/in^2K (mW/mm^2K). For standard board materials without special heat pipes etc., α is in the range 3.87 to 11.6 mW/in^2K (0.006 to 0.018 mW/mm^2K).

These calculations only determine the increase in board temperature due to static currents. To this must be added any large direct, alternating or switched currents causing significant component dissipation. In the past, designers have erred on the generous side, with wide tracks and large boards. With the growing demand for miniaturization, this cannot continue.

Fortunately, modern computer aided design methods are starting to allow thermal analysis to be undertaken quickly and accurately. Using the analogy between current flow and heat flow, any analog electrical simulation program, such as SPICE, can model the static and dynamic thermal considerations (Haskard 1988; Sinnadurai 1985; ISHM 1984). When large thermal systems are being modeled, a virtual memory system is needed but, even then, the time taken to converge can be considerable. More recently, special thermal analysis programs have become available that give near instant response and can plot, in color, the transient thermal response, such as at switch on, so that hot spots and thermal problems can be identified and eliminated (Kallis, Stratton & Bui 1987; Weiss & Langhorst 1988).

Table 10.1 shows the current levels in copper tracks for safe operation. When current levels are less than ½ amp, tracks of 20 mil (½ mm) (or minimum track width allowable if > 20 mil) are acceptable.

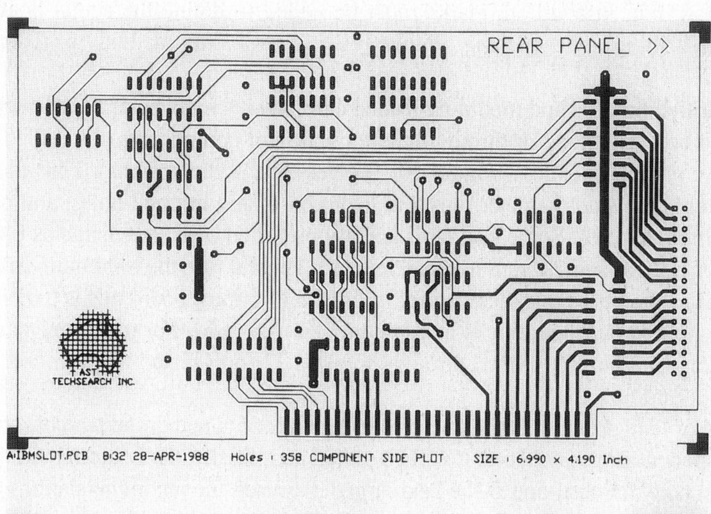

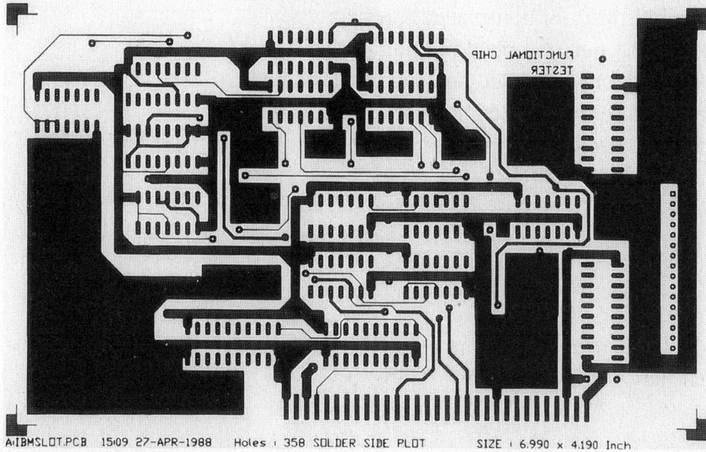

Figure 10.5 Typical method of defining a board size for trimming. Note the connector at one end

(Reproduced with permission of Australian Silicon Technology)

10.2.3 Mechanical considerations

The mechanical considerations include the board dimensions, connections, particularly of edge connectors, pads, vias and test points, drilling information, and device footprints.

Standard board sizes are frequently used in the industry. This allows interchange of cards and the production of a standard range of blank prototyping boards, frames, fittings and modules for packaging systems. The Eurocard is one standard that is commonly employed. Here the basic card sizes are 100 × 160 mm and 100 × 220 mm. Double-size cards 233.4 × 160 mm and 233.4 × 220 mm are also available. The large number of IBM personal computers (including clones) in use has meant that the IBM card and half card sizes have also become standard in much of the industry. However, for many applications other factors such as size, shape, industrial design and cost dictate dimensions. No matter what card size is used the guidelines are the same. All dimensions must be from a datum (Section 10.2), but, in addition, marks must be added to define the actual board size for guillotining, drilling, or punching. Figure 10.5 shows one method to indicate board dimensions.

Holes drilled for vias and mounting leaded components must be of the correct diameter. If the drilled hole is plated, for through-hole mounting of components, the drilled hole must allow for the plating thickness (Figure 10.10 gives typical dimensions). Thus the optimum drill diameter is the lead diameter plus 5 mil for non-plated through-holes, and plus 20 mil for plated through-holes. Some standards specify preferred drill sizes, such as 0.4, 0.5, 0.6, 0.8, 0.9, 1.0, 1.3, 1.6 and 2.0 mm in diameter. As a general rule the minimum drilled hole is about one-third the board thickness, so for a 1⁄16 inch (1.6 mm) board this is 0.02 inches (0.6 mm). If the hole is punched, this is increased to 1⁄32 (0.8 mm). For mounting standard dual in line packages, a No. 68 (plated through-hole) or 72 (non-plated through-hole (0.8mm)) drill size is common.

The pad or land size in relation to the hole depends on whether the land has to support a leaded component or not, whether it will be plated and, finally, the board material used. If d is the drilled hole diameter and D the land or pad diameter, the recommendations are:

$D - d \geq 40$ mil (1 mm) for non-plated holes
≥ 20 mil (0.5 mm) for plated through-holes
$D/d = 2.5$ to 3.0 for non-plated holes in phenolic boards
$= 1.8$ to 3.0 for non-plated holes in epoxy boards
$= 1.5$ to 2.0 for plated through-holes.

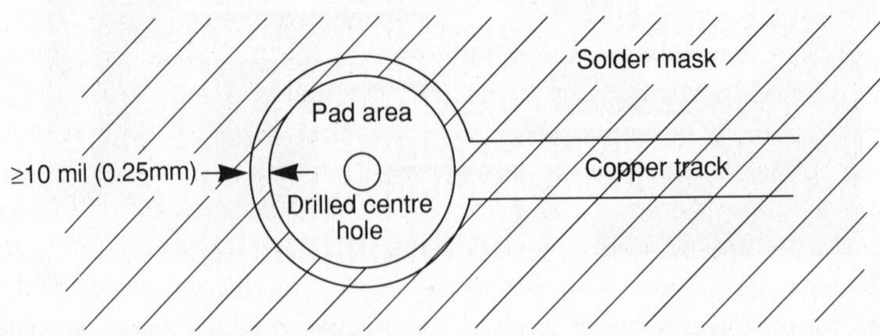

Figure 10.6 Solder mask minimum dimension overlap for a leaded component land

Table 10.2 Typical minimum size test pads and resulting restrictions

Grid size/pitch	Test pad size	Comments
0.1 inch (2.5 mm)	≥ 0.05 inch (1.3 mm)	Standard test probes. Test points on both sides possible, but not preferred.
0.1 inch (2.5 mm)	≥ 0.025 inch (0.7 mm)	Miniature test probes. Test points only one side preferred.
0.04 inch and 0.05 inch (2.5 and 1.0 mm)	≥ 0.025 inch (0.7 mm)	Miniature test probes. Test points only one side of a board.

The solder mask for pads or lands used to mount through-hole leaded components must allow at least 10 mil (0.25 mm) clearance around the pad—that is, the edges of the pad are to be covered by the solder mask. This is illustrated in Figure 10.6.

For convenience, all pads or lands should show their center. The usual method is to omit the center so that the copper is etched away. This aids inspection and assists in drilling of the board with non-plated through-holes. Pads and vias are also used for test points. In the case of a surface mount board, extra probe test points may have to be added. These are normally circular with the center intact. Diameters depend upon the board and grid size. Table 10.2 suggests typical values. It is preferable, if not essential, that test points be placed only on the solder side of a board.

The tracks, pads, vias and test points are laid out on a grid. The preferred grid size was traditionally 100 mil (2.54 mm) but new integrated circuit packages with pin spacings of 50, 40 and 25 mil have undermined this long established standard. Depending on package types, grids of 10, 25, 50 or 100 mil or 0.5 mm may be used. As explained in Chapter 8, the 100 mil grid is much preferred for test points because it allows larger, more robust test pins to be used.

It is important that components and possibly even tracks are not placed too close to the edge of a board because many boards slide in slots in the racks. The minimum safe working distance from the edge of a board for a track is 40 mil (1 mm).

The preferred method of interconnecting pads and vias is to use orthogonal tracks of constant width with 45° angled corners. Track width and spacing greater than or equal to 20 mil (0.5 mm) is acceptable, but in most instances 25 mil (0.625 mm) and 40 mil (1 mm) tracks and spaces are adequate. Many board manufacturers prefer the latter.

For through-hole mounted integrated circuit packages, connectors and components, footprint dimensions have been standardized for many years. With the recently introduced surface mount components, this is not the case. Not only is there some variance in the optimum pad layout for a component, but there is an increasing number of 'standard' passive component sizes and integrated circuit package types. In addition to the standards that are appearing in official documents, almost every major semiconductor manufacturer has issued a document that covers surface mount technology and package footprints and pads (Philips 1986, 1987; Siemens 1986; Mullen 1984; Mullard 1983; AVX 1987; EDC 1987). Such is the competition to have one's package outline and ideas become a standard for the industry (Palmer 1991).

Table 10.3 Rectangular chip component dimensions and recommended land sizes. Dimensions are in inches and mm in brackets. Refer to Figure 10.7

Component	Style	Dimensions			Land size		
		L	W	t	A	X	Y
Resistor	RC 0805	0.070–0.087 (1.8–2.2)	0.040–0.055 (1.0–1.4)	0.012–0.028 (0.3–0.7)	0.020 (0.8)	0.055 (1.4)	0.060 (1.5)
	RC 1206	0.118–0.134 (3.0–3.4)	0.055–0.070 (1.4–1.8)	0.016–0.028 (0.4–0.7)	0.070 (1.8)	0.065 (1.6)	0.065 (1.6)
	RC 1210	0.118–0.134 (3.0–3.4)	0.090–0.106 (2.3–2.7)	0.016–0.028 (0.4–0.7)	0.070 (1.8)	0.100 (2.6)	0.065 (1.6)
Capacitor	CC 0805	0.070–0.087 (1.8–2.2)	0.040–0.055 (1.0–1.4)	≤ 0.05 (≤ 1.3)	0.020 (0.8)	0.055 (1.4)	0.060 (1.5)
	CC 1206	0.118–0.134 (3.0–3.4)	0.055–0.070 (1.4–1.8)	≤ 0.06 (≤ 1.5)	0.070 (1.8)	0.065 (1.6)	0.065 (1.6)
	CC 1210	0.118–0.134 (3.0–3.4)	0.090–0.106 (2.3–2.7)	≤ 0.067 (≤ 1.7)	0.070 (1.8)	0.100 (2.6)	0.070 (1.8)
	CC 1812	0.166–0.190 (4.2–4.8)	0.118–0.134 (3.0–3.4)	≤ 0.067 (≤ 1.7)	0.125 (3.2)	0.125 (3.2)	0.070 (1.8)
	CC 1825	0.166–0.190 (4.2–4.8)	0.236–0.268 (6.0–6.8)	≤ 0.067 (≤ 1.7)	0.125 (3.2)	0.260 (6.6)	0.070 (1.8)

Table 10.4 Dimensions for small outline and plastic J lead package lands. Dimensions are in inches with mm in brackets. Refer to Figure 10.8

Package style	No. of pins	Dimensions			Land size	Land spacing
		A	B	C		
Small outline	8	0.14 (3.6)	0.30 (7.6)	0.15 (3.8)	0.025 × 0.080 (0.63 × 2)	0.05 (1.27)
	14	0.14 (3.6)	0.30 (7.6)	0.30 (7.6)	0.025 × 0.080 (0.63 × 2)	0.05 (1.27)
	16	0.14 (3.6)	0.30 (7.6)	0.35 (8.9)	0.025 × 0.080 (0.63 × 2)	0.05 (1.27)
Plastic J lead chip carrier	28	0.34 (8.6)	0.50 (12.6)	0.42 (10.6)	0.025 × 0.080 (0.63 × 2)	0.05 (1.27)
	44	0.54 (13.6)	0.70 (17.6)	0.62 (15.6)	0.025 × 0.080 (0.63 × 2)	0.05 (1.27)
	68	0.84 (21.2)	1.00 (25.2)	0.92 (23.2)	0.025 × 0.080 (0.63 × 2)	0.05 (1.27)
	84	1.04 (26.4)	1.20 (30.4)	1.1 (28.4)	0.025 × 0.080 (0.63 × 2)	0.05 (1.27)

Figure 10.7 shows the outline for rectangular chip components and the land pattern, while Table 10.3 provides actual figures for standard resistor and capacitor type. Inclusion of vias in SMD pads can provide increased routing space (Braun 1990).

Two integrated circuit packages that are commonly used in surface mount assembly are the small outline (SO, SOP or SOIC) and the J lead chip carriers. Lands for these packages are shown in Figure 10.8 and Table 10.4.

Overglaze patterns vary considerably from one designer/manufacturer to another. Some place the glaze between lands, while others leave the whole land area clear of solder mask. Where tracks to large lands are not covered by a solder mask, these tracks should not be made too fine because, on soldering, the solder shrinks and the fine track either lifts or cracks. The track should be fanned slightly to the pad or covered with the solder glaze.

The orientation of dual in-line and small outline packages needs to be considered when wave soldering is used. Most boards come out from the solder wave bowed and, to minimize stress on the integrated circuit leads, components should be mounted parallel to the longer dimension of the board. Further, when wave soldering is used, care must be taken so that components do not 'shadow' each other and prevent correct soldering.

10.3 Manual layouts

When manual layouts are undertaken temperature stable plastic translucent material such as polyester is used for the artwork. A light box with a translucent grid sheet fixed to the upper surface is used so that the designer has a clear view of the grid. The artwork is produced by fixing black crepe paper tapes, specially prepared with a sticky backing, to the polyester sheet. The tapes come in reels of various widths and a scalpel is used to cut the tape to the correct length. Pads of various diameters and component footprints of various standard shapes and sizes are available.

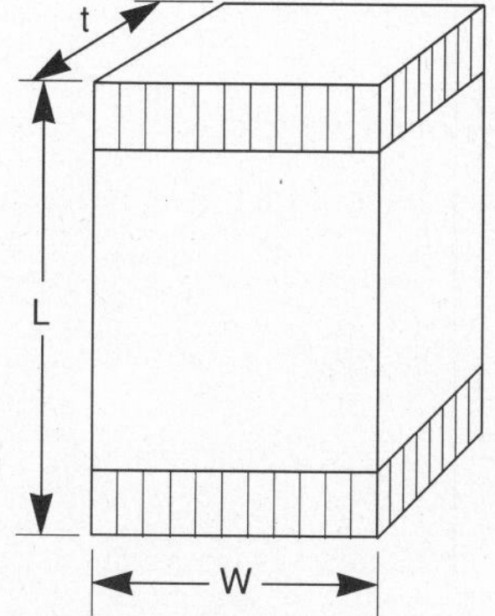

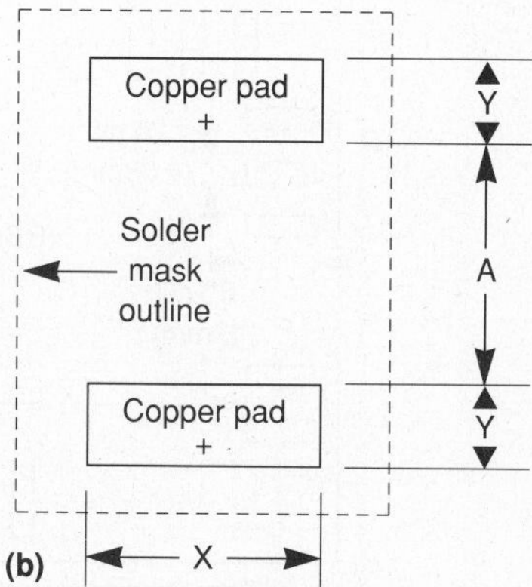

Figure 10.7 (a) Rectangular chip component, and (b) recommended land pattern

Three approaches can be used. Take, for example, a double-sided plated through-hole board. The normal approach is to produce taped artwork for each side of the board, each with tracks and pads. Care must be taken that pads on each board side are accurately aligned. A variation on this is to have only the tracks on these artwork sheets and the pads placed on a separate third sheet. This guarantees alignment. The third method is to produce a single drawing with pads in black crepe and the tracks for either side of the board in different colors: blue and red. During the photographic reduction stage filters are used to separate the two layers.

Artwork is normally done at × 2; however, where additional precision is needed, × 4 or × 10 can be used.

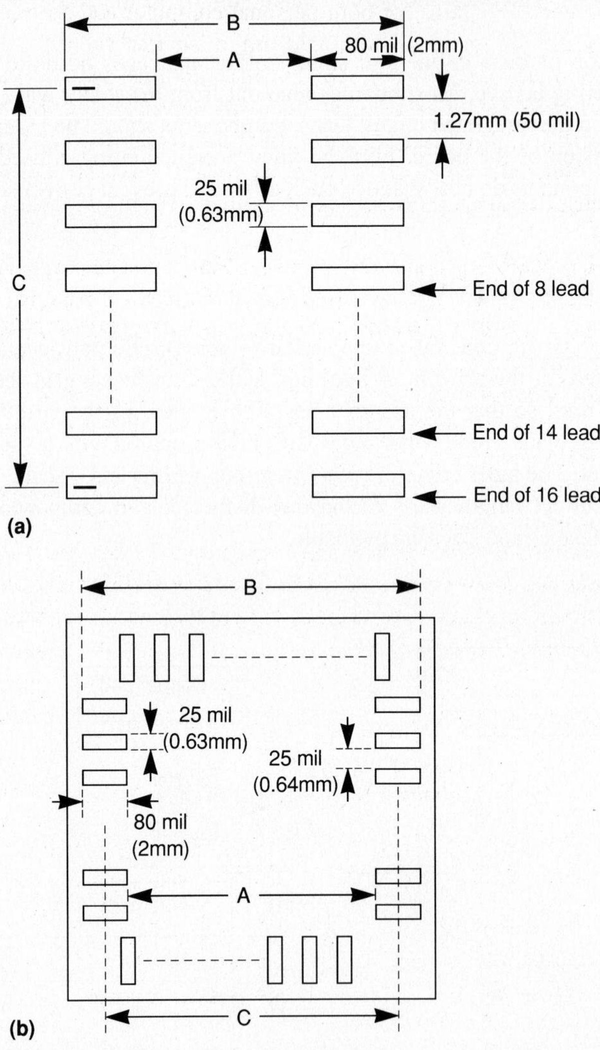

Figure 10.8 Recommended land layouts for (a) small outline and (b) plastic J lead chip carrier integrated circuit packages

The artwork must give alignment or datum marks and board edge trimming dimensions. Figure 10.10 illustrates one method and gives simple rules for manual layout of a double-sided, plated through-hole board.

In addition to this artwork, solder glaze and screen printing of component types must be generated. Once this has been produced, mechanical drawings and drill information can be documented.

10.4 Computer aided design methods

In recent years there has been a rapid growth in the availability of software for the design of printed circuit cards. Programs for both personal computer and more expensive work stations are available. In general, both perform a similar function except that the workstation versions tend to have more options, larger libraries and operate much faster.

Early programs were simple geometric editors allowing the placement and routing of tracks. Being interactive, it was easy to erase, shift and replicate components and blocks of circuitry. Unfortunately some of the early software was only suitable for leaded components because it did not allow grids other than 100 mil or components on both sides of the board.

With the passage of time, schematic entry was added to the geometric editor which, in turn, allowed automatic routing. Today automatic placement and routing is possible. The software library now contains not only integrated circuit standard footprints, but full electrical data on standard product lines (TTL and CMOS packages at least) so that simulation can be undertaken on digital circuits, after the software has extracted all relevant parasitics and added them to the circuit. Net lists, drill sizes and other relevant facts are automatically generated.

As explained in Section 10.2.2, thermal analysis programs now allow not only electrical, but thermal simulation.

The software is normally written to enable each program to function independently of the other, so users need only purchase those that are required for a particular job. Programs do operate together as a total integrated system, however, with pop-up menus/windows listing a range of operations which can be selected with a mouse.

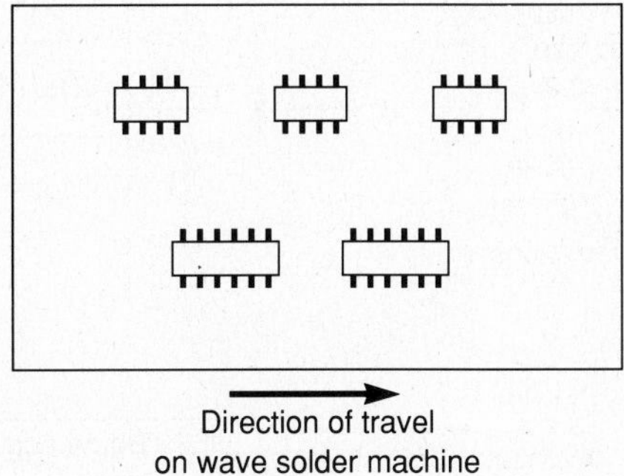

Figure 10.9 Direction of mounting of small outline packages on a board for wave soldering

Selection of an appropriate computer aided design system is difficult and many articles have been published to assist people (Warren 1989; Zibaldo 1989; Audeh 1989; Simon 1989; Freeman 1987). Summary charts with comments are also available (*Computer Design* 1988). The early, simple programs that ran on personal computers have now become quite sophisticated. Examples of such programs are PROTEL, SMARTWORK, P-CAD and CAD-STAR.

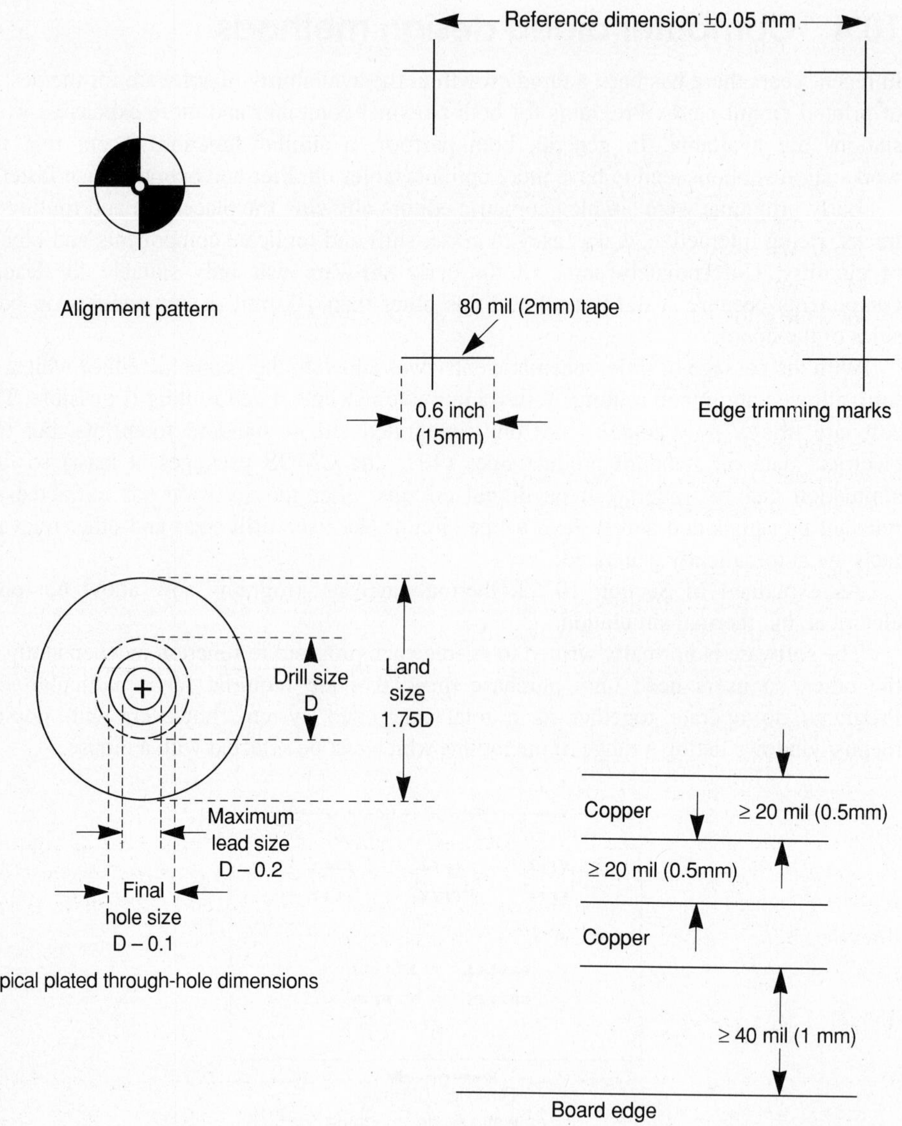

Figure 10.10 Simple layout rules for a double-sided, plated through-hole printed circuit boards. Imperial units given first with metric in brackets

If designing printed circuit cards is not your full time occupation, it is often better to obtain a simple system which has minor restrictions. The instruction set should be easy to recall so that overall productivity can be higher. Factors that should be considered include:

1. General
   Type and memory capability of computer required
   Range of software
   Size of data base/library
   Local support

2. Specific
   Schematic entry
   Manual/automatic placement (efficiency)
   Manual/automatic routing (efficiency)
   Multilayer capability (maximum number of layers)
   Zoom and pan
   Maximum board size
   Variable grids
   Net list generation
   Test vector generation
   Thermal analysis
   Screen display style
   Maximum number of components/nets
   On-board ruler/measuring device
   Highlight tree
   Rubber banding
   Surface mount capability
   Miter corner
   Automatic component name/rename
   Test justification
   Test plots (HP format and others)
   Photo plotter (Gerber format and others)
   Components placed on both sides

Above all, the software must be capable of producing all the desired outputs listed in Figure 10.1, be user friendly and capable of handling the complexity of the boards which will be designed over the next 3–5 years. The artwork reproduced in Appendix B was drawn using the simple to use PROTEL CAD package.

10.5 Questions

1. Refer to Figure 10.1. There are seven outputs listed from the PCB layout and design phase. Explain why each is required. Which is software, artwork and/or documentation?
2. An essential component of PCB documentation standards is a method of recording changes so that everything is systematically and correctly updated. Either find out how

this is achieved in your place of employment or devise a system of your own that is simple to implement.

3. Obtain from a supplier a catalog of Eurocard parts. Study it carefully. Assuming the power supply (excluding the power transformer, meter, potentiometer, fuse, terminals, and switches) of Figure B.5 has been constructed on a standard Eurocard, select:
   (a) a suitable case for a self contained mains operated power supply
   (b) frames and panels to construct a modular case suitable for an instrument where the power supply feeds a further three Eurocards, all to be mounted in the same case. (Here the power supply does not need to have terminals brought out to the front panel, but the meter, switches and voltage set controls are needed. The additional front panel instrument controls are a 3½ digit liquid crystal display, 6-position range switch and input signal connector.)
4. When laying out a card, what important factors must be considered in the following areas?
   (a) mechanical design
   (b) electrical design
   (c) thermal design
5. 'The layout of a surface mount board is more difficult than a through-hole leaded component board.' Discuss this statement, listing the reasons for and against.
6. Complex surface mount boards often have more holes than leaded through-hole boards. Why?
7. You are designing a double-sided, plated through-hole board. List the layers for which you must prepare artwork.
8. Using the simple rules given in Figure 10.10, rough out a layout at ×10 for the circuit given in Figure 10.11 using:
   (a) leaded components (8 pin DIL package)
   (b) surface mount components (8 pin SOP package).
   (Use the typical dimensions given in Section 10.2.3.)

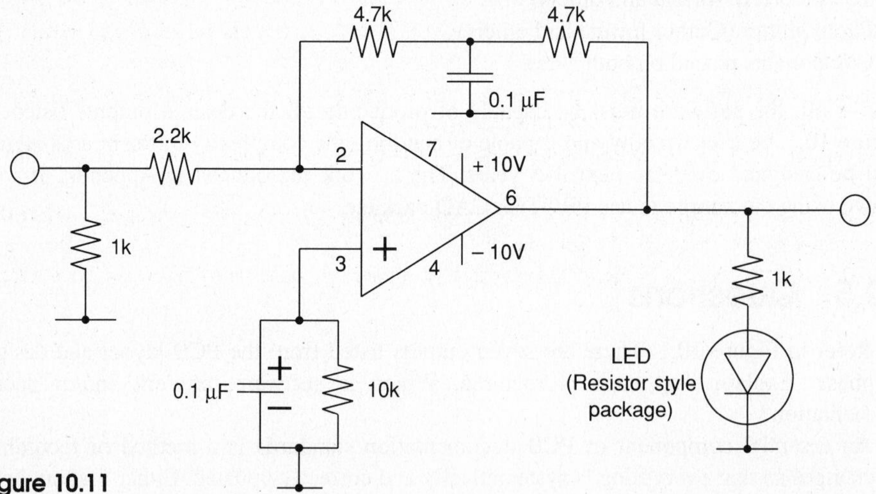

Figure 10.11

9. Repeat the exercise in Question 8 using a CAD package and typical components from the software package library (for example, a 741 operational amplifier). How compact can you make the layout?

10.6 References

AVX (1987), *Surface Mounting Guide,* AVX Ltd, Aldershot. UK.

Audeh, A. S. (1989), 'Tools of the trade', *Printed Circuit Design*, Vol. 6, (March), pp.13–17.

Barrett, R. & De Maria, D. (1987), 'Bench-marking PCB design systems', *Computer Design*, **26** (1 January), pp. 77–82.

Braun, R. (1990), 'A unique way of placing via hole connections in the SMD pads gives increased routing space on printed circuit boards', *Circuit World*, **16**(2), pp. 45–6.

Computer Design (1988), 'Designers' Buying Guide', **27** (1 June), pp. 87–143.

EDC (1987), *Introducing surface mounting*, EDC Electronic Components, London.

Freeman, E. (1987), 'Low cost PC board layout software', *Ham Radio*, **20**(10), pp. 8–15.

Haskard, M. R. (1988), *Thick Film Hybrids: Manufacture and Design*, Prentice Hall, New York.

ISHM (1984), *Thermal Management Concepts in Microelectronic Packaging*, International Society of Hybrid Microelectronics (Technical Monograph 6984–003), Silver Springs.

Kallis, J. M., Stratton, L. A. & Bui T. T. (1987), 'Programs help spot hot spots', *IEEE Spectrum*, **24** (March), pp. 36–41.

Mullard (1983), *Surface Mount Devices*, Philips, London.

Mullen J. (1984), *How to Use Surface Mount Technology*, Texas Instruments.

Palmer, E. G. (1991), 'Chip Component Pad Dimensioning', *Printed Circuit Design*, **8**(5), pp. 32–4.

Philips (1986), *SMD Technology*, Philips Electronic Components and Materials, Eindhoven, Netherlands.

Philips (1987), *SMD Packaging Systems: Total Expertise in SMD Technology*, Philips Electronic Components and Materials, Eindhoven, Netherlands.

Simon, C. J. (1989), 'Evaluating CAD Systems', *Printed Circuit Design*, **6** (March), pp. 8–12.

Siemens (1986), *Components for Surface Mounting: PCB Layout Recommendations*, Siemens AG, Munich.

Sinnadurai, F. N. (1985), *Handbook of Microelectronic Packaging and Interconnection Technologies*, Electrochemical Publications, Ayr, Scotland.

Warren, P. (1989), 'The pitfalls of selecting a CAD system', *Printed Circuit Design*, **6** (March), pp. 26–8.

Weiss, J. & Langhorst, F. (1988), 'Thermal analysis of electronic packaging', *Australian Electronics Engineering*, **21** (August), pp. 60–4.

Zibaldo, C. (1989), 'Choosing a CAD system: A different perspective', *Printed Circuit Design*, **6** (March), pp. 18–25.

11 Extensions of the technology

11.1 Introduction

The printed circuit board has been adapted and applied to a number of fields to produce different components. The reasons for this include the ease with which the new components can be combined on the same board.

In this chapter three classes of such components will be considered: printed circuit inductors and transformers, stripline techniques and sensors.

11.2 Printed circuit inductors

Any length of wire has a self inductance (approximately 1 μH/meter). Consequently, circuits requiring small inductors can be manufactured using the copper track. Various geometries can be used, including flat, square spirals, meander and zig zag (Casse 1969; Green 1972). Figure 11.1 shows a printed circuit card used in the front end of a VHF TV set containing flat spiral inductors.

Of the various possible geometries, the flat spiral type is most common because it provides the greatest inductance per unit area. A via is needed so that the center track of

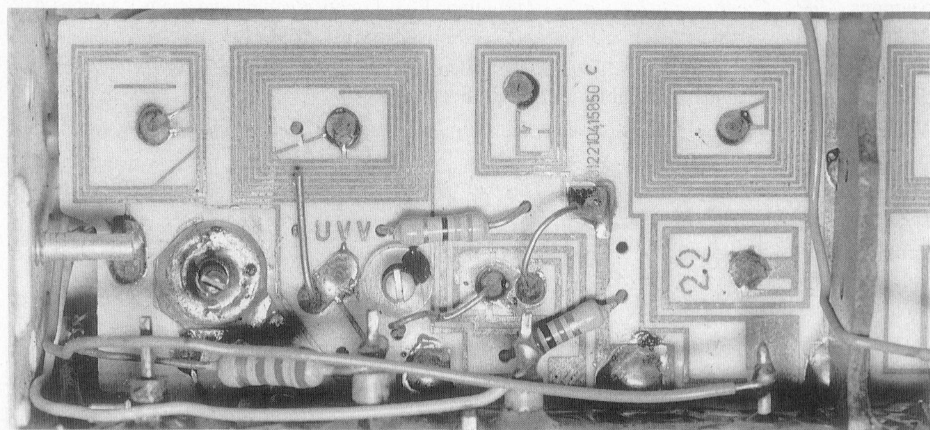

Figure 11.1 Printed flat spiral inductors

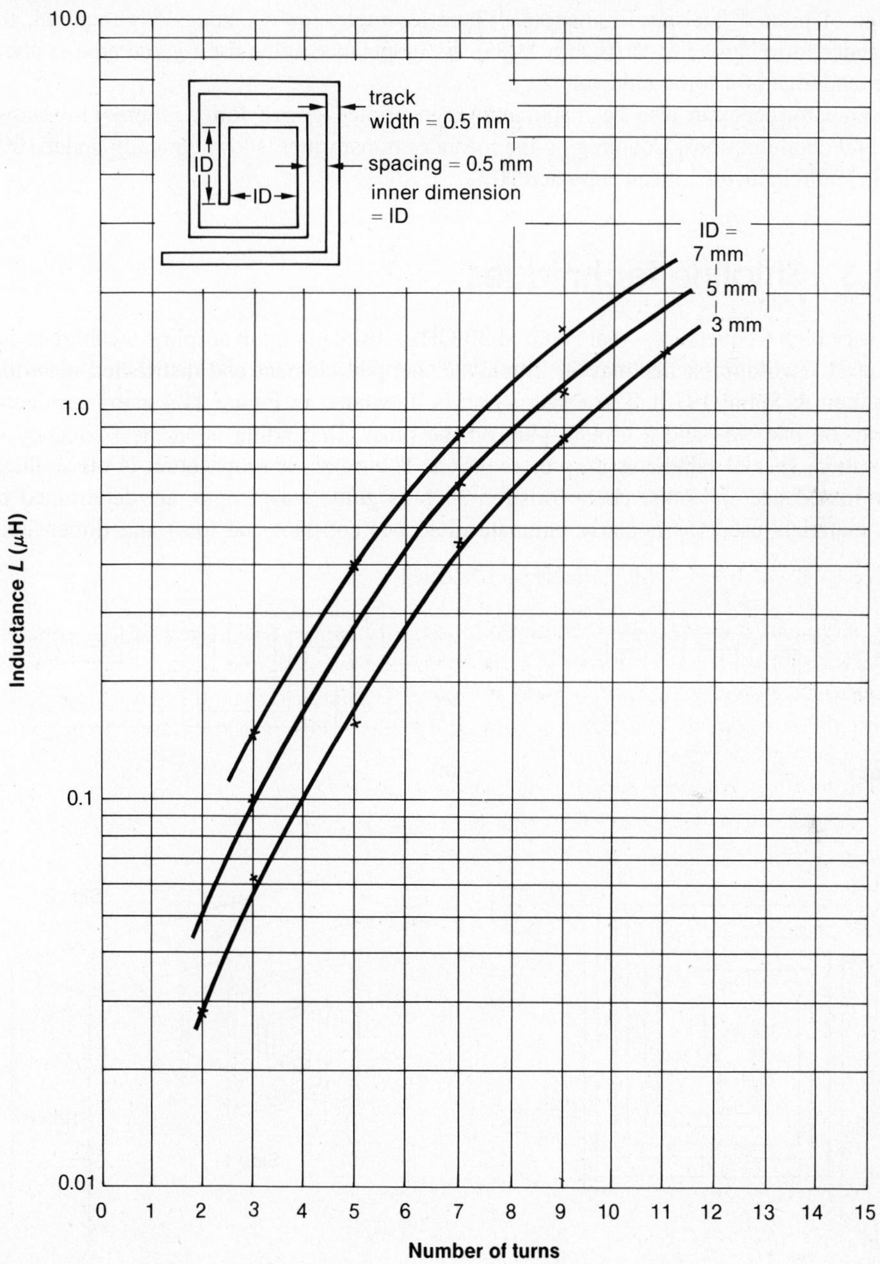

Figure 11.2 Inductance values and shape of printed flat spiral inductors

the conductor can be brought out. Further, with a double-sided board, the inductance per unit area can be increased by placing turns on both sides of the board. Q factors for such inductors can be a 100 plus. Figure 11.2 gives the inductance values and shape for a range of printed flat spiral inductors. Where a simple, low Q inductor is required, the meander form can be used (Figure 11.3a). Its inductance value for a given area is about one tenth that of a spiral inductor.

Transformers can also be manufactured in printed circuit form. Figure 11.3 shows several configurations. Coupling on the meander transformer is low, typically under 10%. With spiral inductors, it can approach 90%.

11.3 Stripline techniques

At very high frequencies—that is, up to 30 GHz—printed circuit stripline techniques can be used. Two approaches may be employed, lumped element and distributed networks (Caulton & Sobol 1970). The construction is illustrated in Figure 11.4 using conductor tracks on one side and a ground plan on the other. Depending upon the frequency of operation, special laminates may be required. The mode of propagation is quasi TEM. The impedance and other characteristics, such as guide wavelength, are determined by the materials used (particularly laminate dielectric constant and loss) and dimensions.

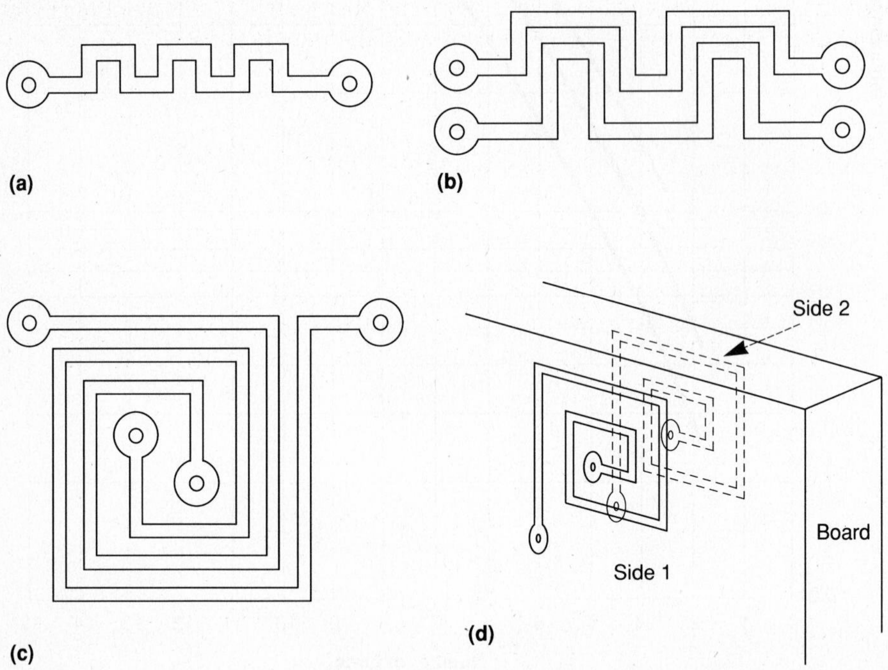

Figure 11.3 Examples of printed circuit inductors and transformers: (a) meander inductor, (b) meander transformer, (c) spiral transformer on one side of a board, and (d) using two sides of a board

The important dimensions are W the upper track width, H the laminate thickness and G the width of the ground plane. These are shown in Figure 11.4. Ideally G should be infinite, but in practice it can be reduced to 10 W. For very low impedances ($Z_0 \leq 30$ ohms) it can be reduced even further to 5 W.

The design of circuits using stripline/micro-strip circuits will not be considered in this text (see Fooks & Zakarevičius 1990; Edwards 1981), but rather the method of constructing the circuit. Two methods are used, one employing self-adhesive copper tape (Kristiansen 1988) and the other, conventional photolithography (Atkins 1988).

In the case of the self-adhesive tape method, the copper tape, which is purchased on reels in widths up to 1 inch (25 mm), is cut to the calculated dimensions and pressed onto the appropriate bare laminate. The circuits can be trimmed by shortening or lengthening the piece of stripline. If care is taken in cutting and placement, the circuit performance should be within 15% of the theoretical calculations. Although this method is not suitable for mass production, it is ideal for prototyping and design research. The second method employs conventional printed circuit board techniques. Even standard laminates can be used up to one or two GHz frequency, thereafter special low loss microwave laminates must be employed. Wheeler (1965) has calculated expressions for stripline parameters. Curves showing characteristic impedance for various common dielectric materials are shown in Figure 11.5.

While straight lines are preferable, it is often necessary to use bends. For frequencies up to several GHz, low VSWRs are achieved if significant bends are trimmed at 45°, as shown in Figure 11.6a. Coaxial connectors to the boards are best inserted through the ground plane rather than at the board edges. This is illustrated in Figure 11.6b.

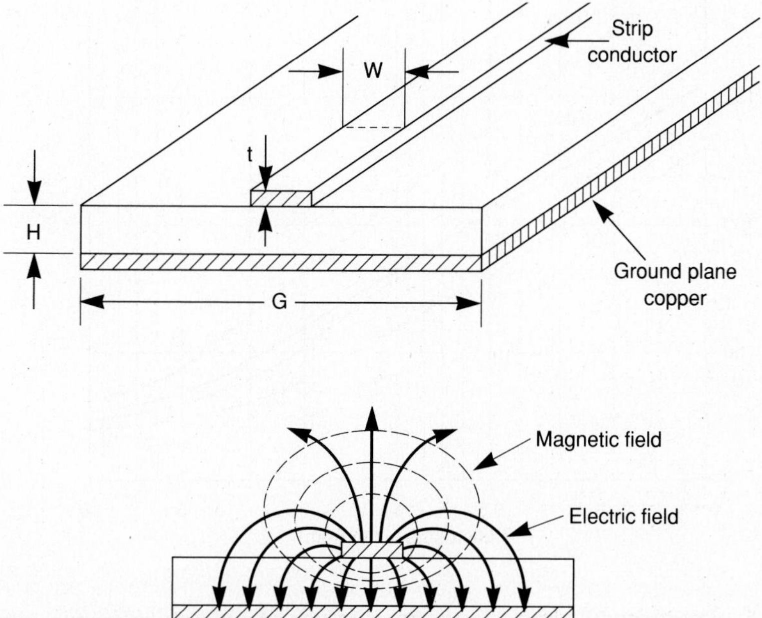

Figure 11.4 Stripline construction

11.4 Sensors

When discussing sensor design today, most people think of silicon and film technologies. While these technologies have much to offer, printed circuit board technology should not be overlooked. In many cases it offers a cost effective solution. Further, it is readily available and allows direct integration into electronic circuitry. Several printed circuit board sensor examples are given below. They are normally based on resistance and capacitive type sensors.

11.4.1 Capacitive based sensors

Figure 11.7 shows one principle of construction. An active capacitor plate is mounted on the board, while the second plate is earthed and variable. By sensing the capacitance change, a range of physical parameters may be monitored (Haskard 1986). Differential capacitors may also be used.

Extremely small displacements may be detected by using a series of printed circuit board fingers fed by voltages of different phases, typically 0, 90, 180 and 270 degrees

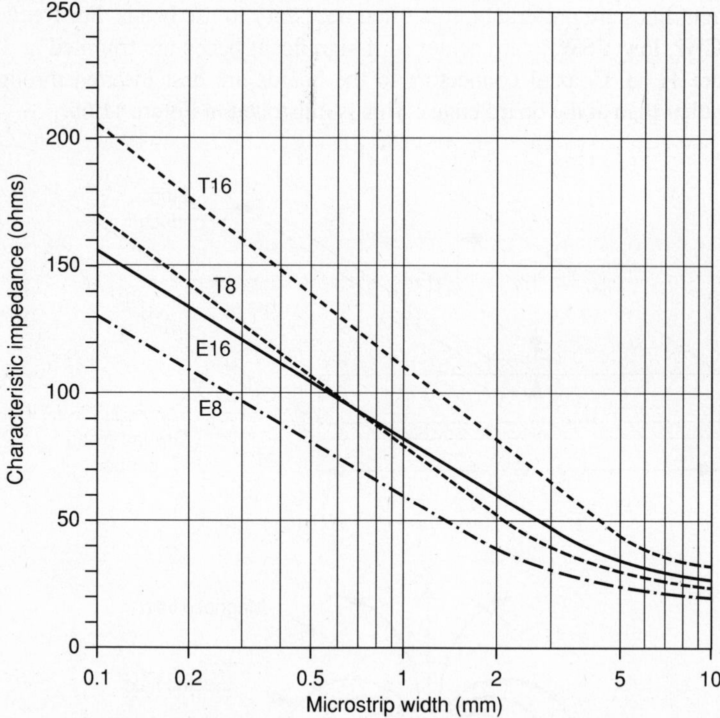

Figure 11.5 Characteristic impedance for striplines of various widths and laminate materials: Teflon laminate, dielectric constant 2.5, 1.6 mm thick (T16), 0.8 mm thick (T8); Epoxy laminate, dielectric constant 5, 1.6 mm thick (E16), 0.8 mm thick (E8)

phase shift (Klaassen, Van Peppen & Sandbergen 1980). This is illustrated in Figure 11.8. When a flat capacitor plate moves over the fingers, the voltage induced will have a resultant phase dependent upon its position.

11.4.2 Resistive based sensors

Simple moisture meters can be made from gold-plated fingers. Normally the resistance between the fingers is very high, but once any moisture has condensed on the fingers the resistance decreases significantly. Special coatings such as metal phthalocyaines can be used to make gas sensors or polyvinyl acetate (PVA) humidity sensors (Bannigan, Friebel, Haskard & Mulcahy 1991).

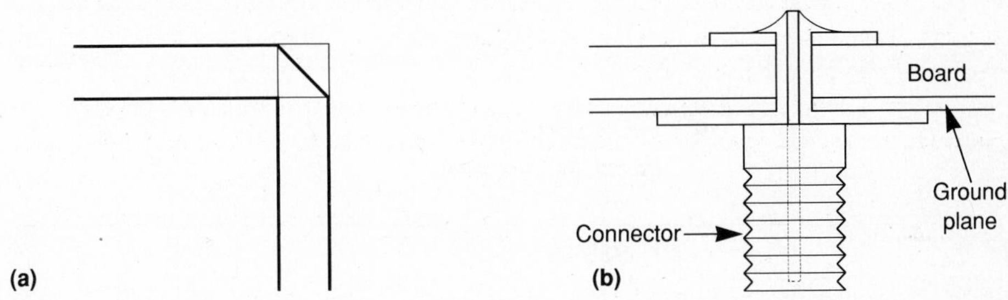

Figure 11.6 Stripline techniques: (a) trimming the corner to 45° to maintain a good VSWR, and (b) mounting of an RF connector

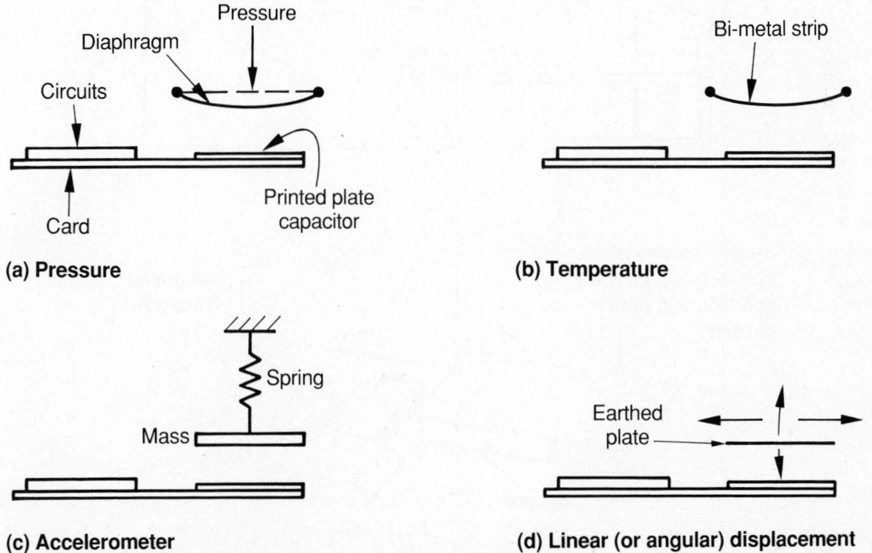

Figure 11.7 Examples of a capacitor sensor

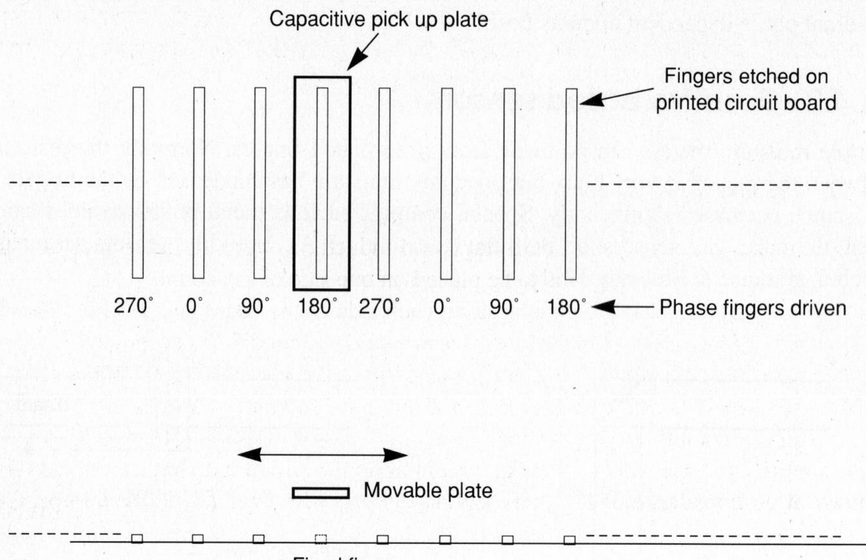

Figure 11.8 Capacitive displacement sensor

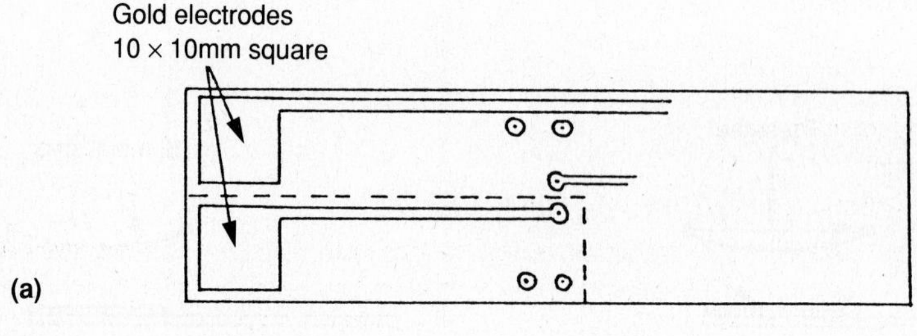

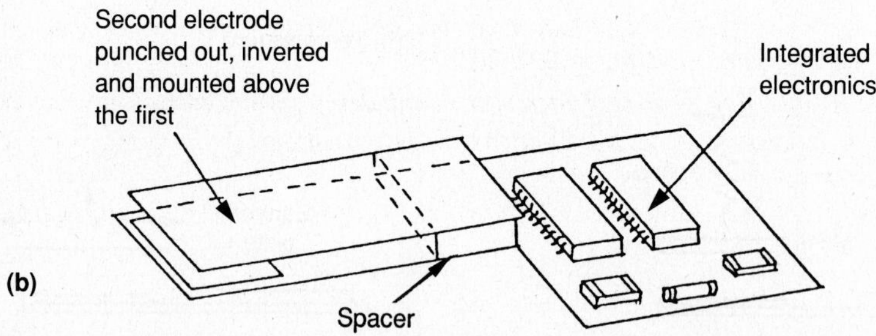

Figure 11.9 Conductivity meter based on printed circuit board technology. Figure shows the initial printed circuit card

Figure 11.9 shows a conductivity cell designed to measure water salinity levels. It uses standard gold plated electrodes, with the second electrode punched from the parent board and mounted at a specified distance above the first by using a spacer and several through-hole soldered leads. On-board electronics allow the manufacture of an inexpensive instrument (Jarmyn, Haskard & Mulcahy 1987).

11.5 Questions

1. (a) Design a printed circuit board flat spiral inductor to have an inductance of 1 μH. Assume that the inductor is to be placed on one side of the board.
   (b) If half of the inductance is placed on each side of the board and, assuming a unity coupling factor between each half, how would you modify your design?
2. An epoxy laminate board 1⁄16 inch (1.6 mm) thick is to be used to make a 1 GHz RF amplifier. The characteristic impedance of the input and output lines is to be 50 ohms. Determine the width of the stripline.
3. A stripline low pass filter is to be incorporated in the circuit with inductors made from lines of an impedance of 80 ohms and capacitors from lines of 30 ohms impedance. What width would you make these lines?
4. Several printed circuit board sensor examples have been described. Can you think of other PCB sensor examples based on:

   (a) capacitance effects?
   (b) resistance effects?
   (c) any other effect?

11.6 References

Atkins, B. (1988), 'Microstripline circuitry', *QST*, **72**(6), pp. 80, 54.

Bannigan, J. T., Friebel, M., Haskard, M. R. & Mulcahy, D. E. (1991), 'Conductivity sensors mode using hybrid technology', *Conference Digest (Microelectronics '91 Conference)*, Institute of Engineers, Melbourne, p. 99–100.

Casse J. L. (1969), 'Printed transformers for high frequency', *Electronic Engineering*, **41** (June), pp. 34–8.

Caulton, M. & Sobol, H. (1970), 'Microwave integrated circuit technology: A survey', *IEEE Journal Solid State Circuits*, **SC-5**(6), pp. 292–303.

Edwards, T. C. (1981), *Foundations of microstrip circuit design*, Wiley Interscience, Chichester.

Fooks, E. H. & Zakarevičius, R. A. (1990), *Microwave Engineering Using Microstrip Circuits*, Prentice Hall, Sydney.

Green, K. (1972), 'Design curves for flat square spiral inductors', *Electronic Components*, **11** (February), pp. 121–6.

Haskard, M. R. (1986), 'General purpose intelligent sensors', *Microelectronics Journal*, **17**(5), pp. 9–14.

Jarmyn, M., Haskard, M. R. & Mulcahy D. E. (1987), 'The design of portable conductivity meters for agricultural applications', *Australian Journal of Instrumentation and Control*, **2**(3), pp. 18–21.

Klaassen, K. B., Van Peppen, J. C. L. & Sandbergen, R. E. (1980), 'Thin-film micro-displacement transducer', in W. A. Kasier & W. E. Proebster (eds), *Electronics to Microelectronics*, North Holland Publishing Company, Amsterdam.

Kristiansen, J. (1988), 'Using copper tape to fabricate RF breadboard circuits', *Hewlett Packard Bench Briefs*, 1st Quarter, pp. 1–4.

Kristiansen, J. (1989), 'How to build a stripline filter', *Hewlett Packard Bench Briefs*, 1st Quarter, pp. 1–5.

Wheeler, H. A. (1965), 'Transmission line properties of parallel strips separated by a dielectric sheet', *IEEE Transaction Microwave Theory*, **MTT-13**, pp. 172–85.

APPENDIX A: Soldering practice boards

Artwork is provided for four different solder practice boards. They are:

Figure A.1 Leaded component practice (using tinned copper wire).
Figure A.2 Leaded component with different pad sizes, swaging and soldering to bifurcate posts and wire wrap pins.
Figure A.3 Surface mount solder practice.
Figure A.4 Mixture of soldered joints and wire wrap (see also Figure 7.20).

Note: Wire-wrap pins, turret and bifurcate posts can be mounted on any pads of sufficient size.

The ideal size for a drill for single-sided board is the component lead diameter plus 5 mil.

Typical drill sizes for various components:

Tinned copper wire—No. 70 (0.7 mm)
Wire wrap pin—No. 59 (1 mm)
Bifurcate/turret post—No. 51 (1.7 mm)
Resistors (1/4 watt)—No. 68 (0.8 mm)
Semiconductor devices—No. 64 (0.9 mm)
Capacitors—No. 64 (0.9 mm)
Screws for heat sinks/mounting—No. 30 (3.5 mm)

Figures in brackets are nearest metric drill sizes.

The large rectangular copper area allows students to scratch their names on the board for identification.

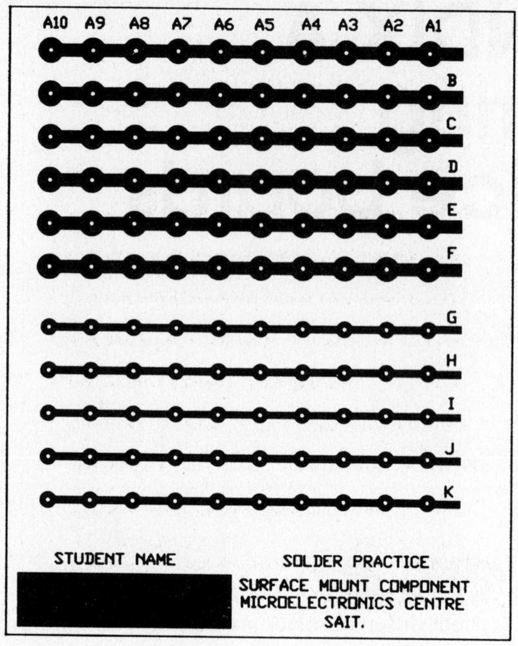

Figure A.1

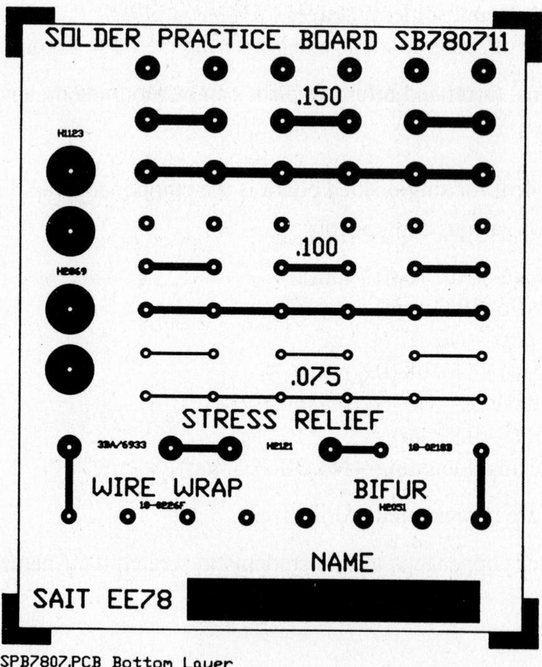

Figure A.2

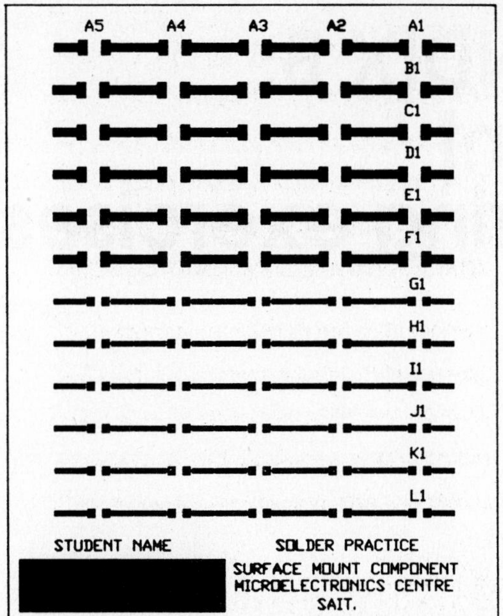

Figure A.3

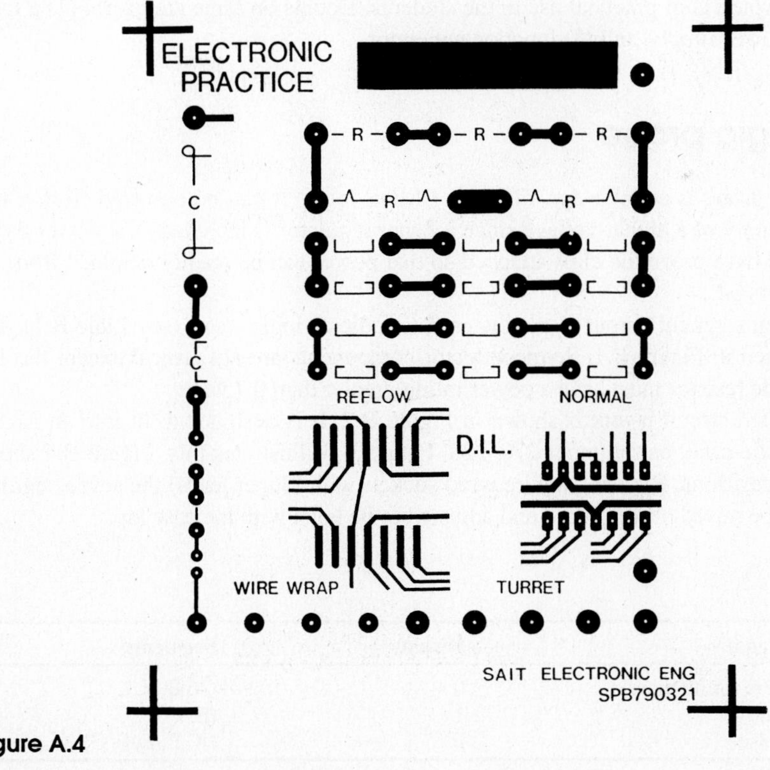

Figure A.4

APPENDIX B: Student soldering exercises

To help students gain skills in soldering and assembly methods, a number of projects have been developed. One of the main objectives was to devise projects that produce equipment which is of practical use to the students. Details on three are given: (1) a logic probe, (2) power supply, and (3) function generator.

B.1 Logic probe

This simple probe is suitable for TTL and CMOS logic. It can be powered from a five volt logic supply or a higher voltage since a Zener regulator is included. The power flying leads should have crocodile clips attached so that power can be readily obtained from the system under test.

The seven segment output display is used to indicate logic states (see Table B.1). The circuit is given in Figure B.1. Semiconductor components are not critical except the 150 Ω zener diode resistor must have a power rating greater than 0.7 watts.

The printed circuit layout is shown in Figure B.2. It is designed to fit into an Archer (Tandy) plastic case, part number 270–220. Figure B.3 illustrates this. Figure B.4 shows component positions. By using a wire wrap socket (with longer leads) the seven segment display can be raised off the board and adjusted to be level with the case top.

Table B.1

Signal	Display	Segments
Zero (or no input)	0	(A, B, F, G)
One	1	(E, F)
Pulse	P	(A, B, E, F, G)

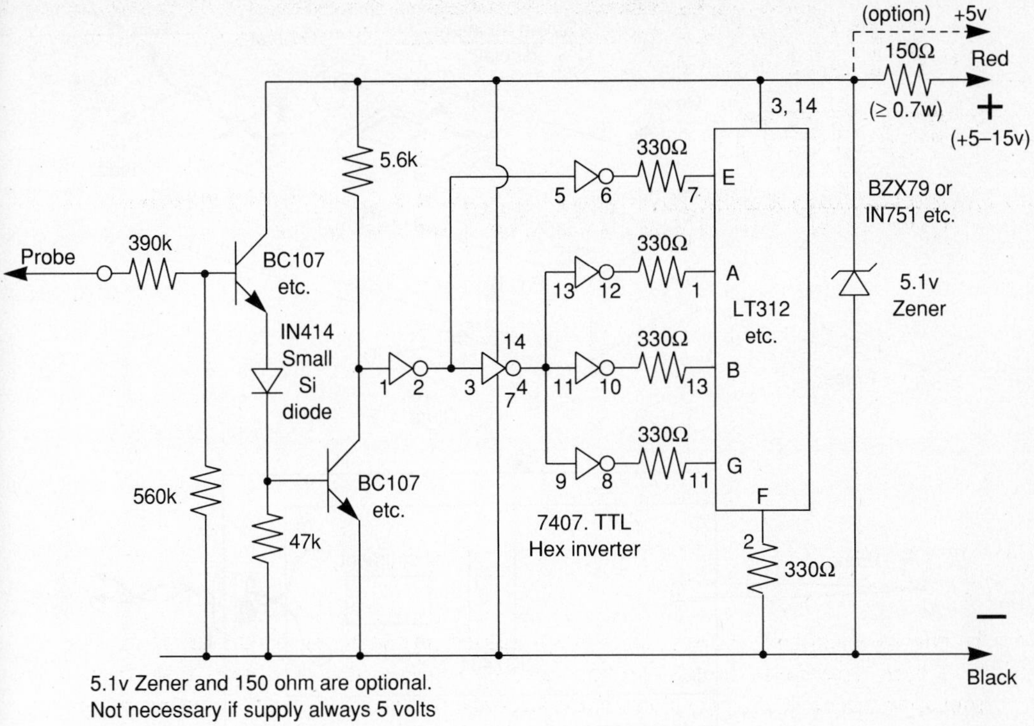

5.1v Zener and 150 ohm are optional.
Not necessary if supply always 5 volts

Figure B.1

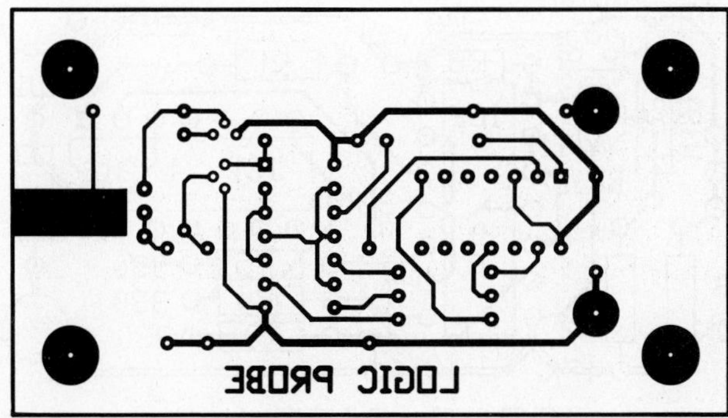

MRHLPROB.PCB Bottom Layer

Figure B.2

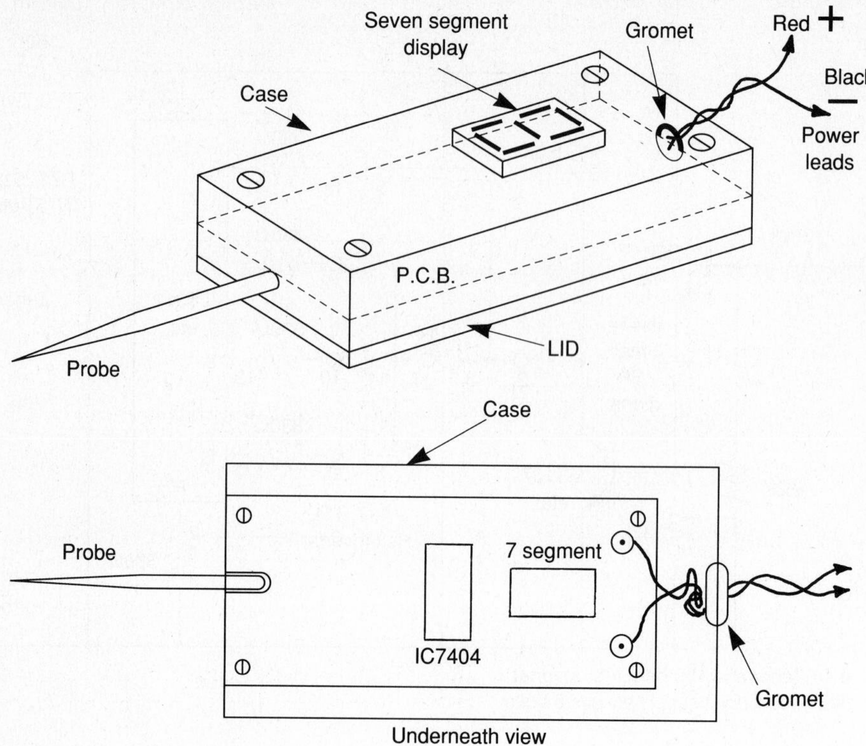

Figure B.3

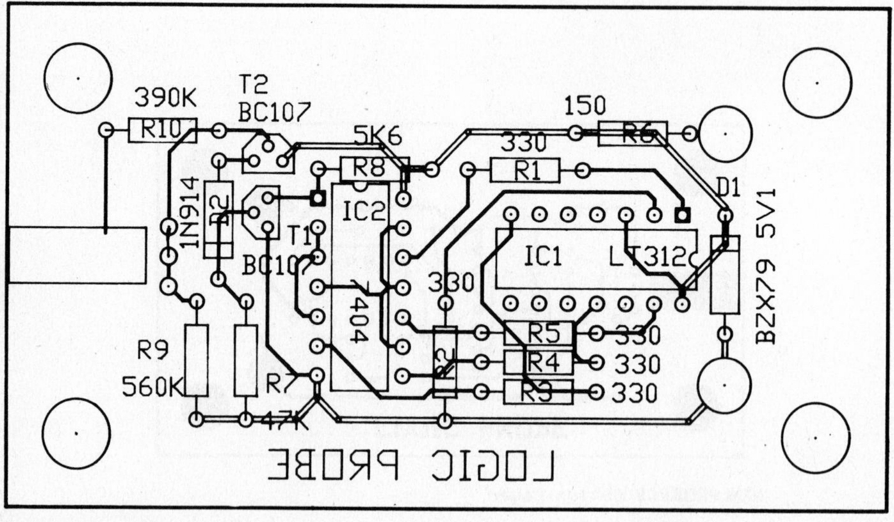

Figure B.4

B.2 Power supply

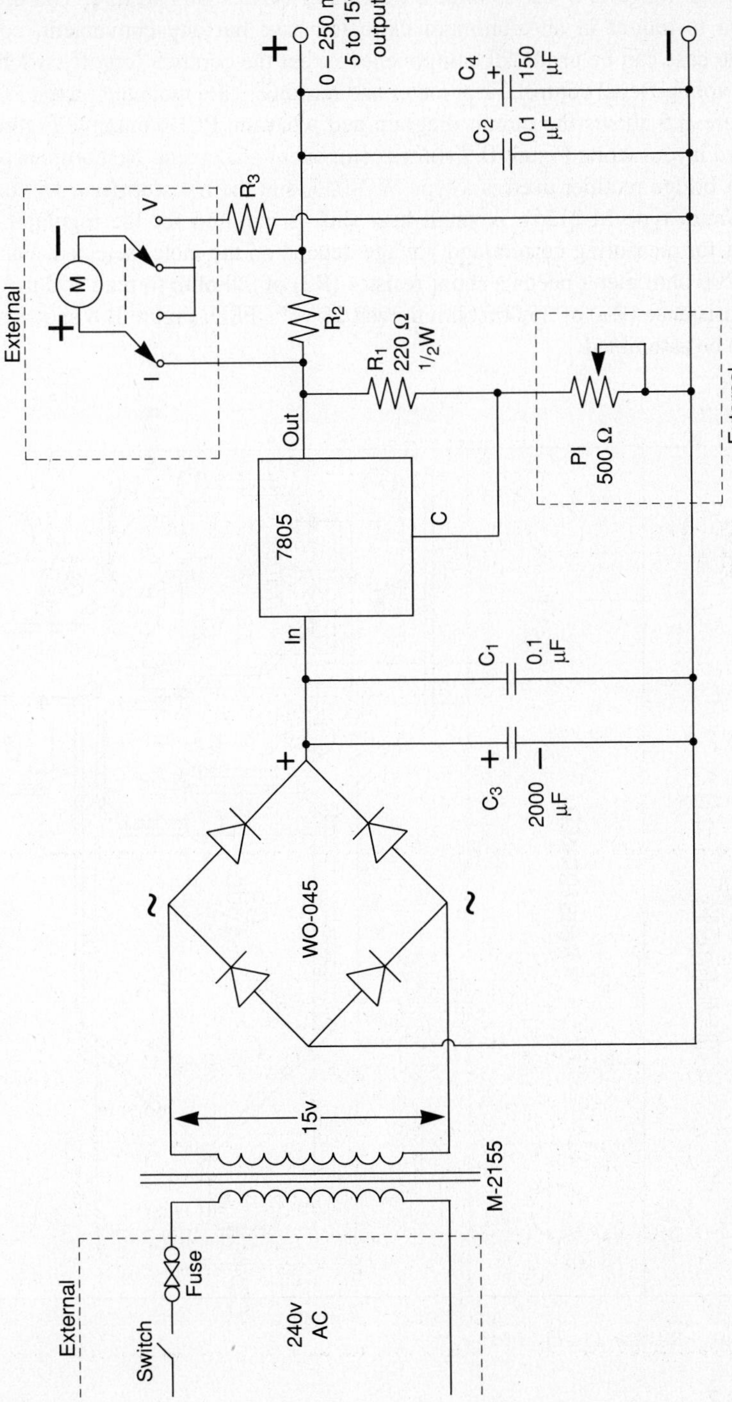

Figure B.5

This is a variable voltage regulated power supply, with meter, based on a 7805 regulator. The voltage range is 5–15 volts at a maximum current of 250 mA. The unit has been designed to mount in an aluminum extrusion case but any convenient, commercially available case can be used. All components except the controls (on/off switch, V/I meter switch, voltage level control, fuse, meter and terminals) are mounted on the PCB.

Figure B.5 shows the circuit diagram and what the PCB contains. Figure B.6 gives the board layout while Figure B.7 shows component placement. No component is critical. The 1 A bridge rectifier used is a type WO-045, and the transformer a 15 volts 1 A type (Dick Smith type M-2155). A small heat sink is required for the regulator. The meter resistors for measuring current and voltage depend on the meter selected. For example, a 1 mA, 200 ohm meter needs a shunt resistor (R2) of 0.8 ohm to read 250 mA FSD and a series resistance (R3) of 25,000 ohm to read 25 volts FSD. Figure B.8 shows one way the unit can be assembled.

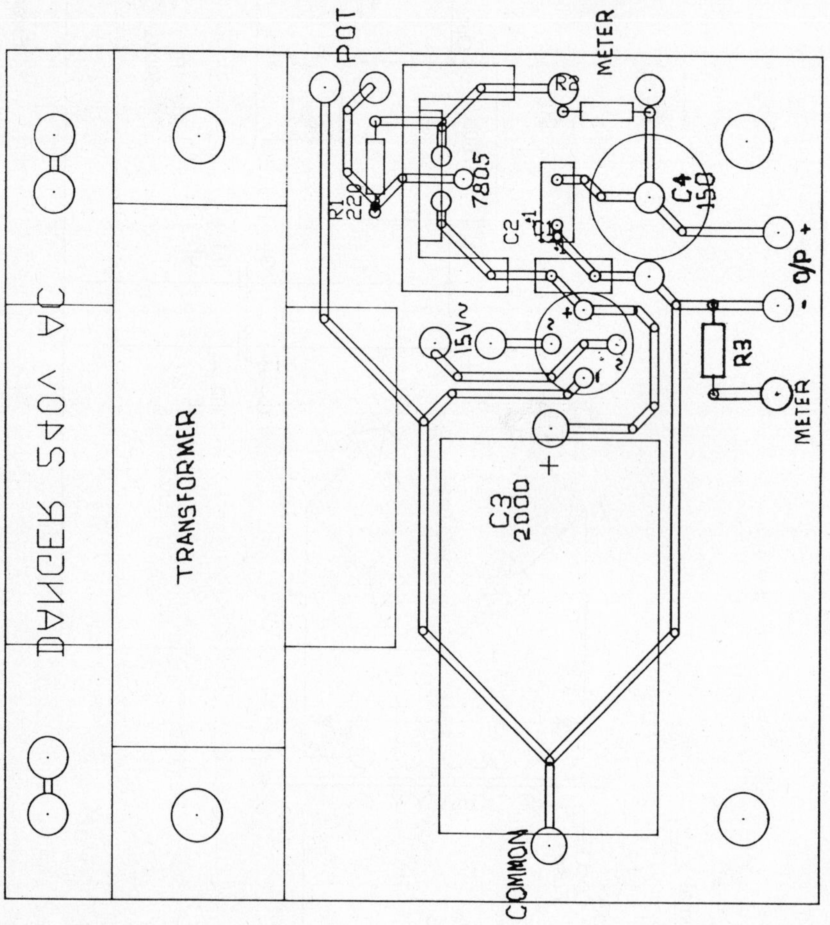

Figure B.7

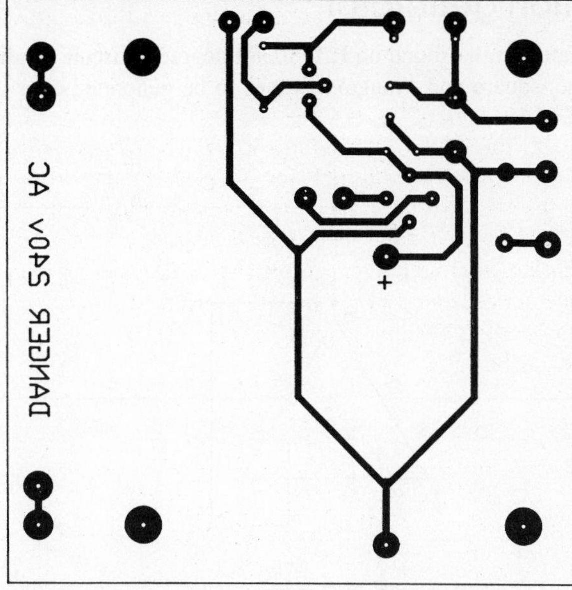

Figure B.6

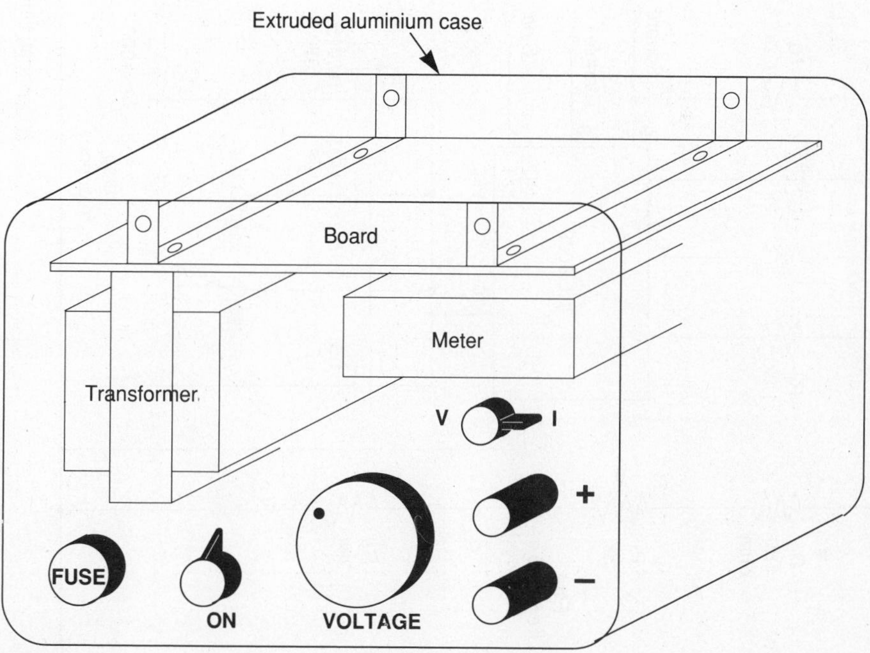

Figure B.8

B.3 Function generator

This small generator, built around an ICL 8038 integrated circuit (Radio Shack part 276-2334) allows sine, square and triangular waves to be generated over the audio range. A

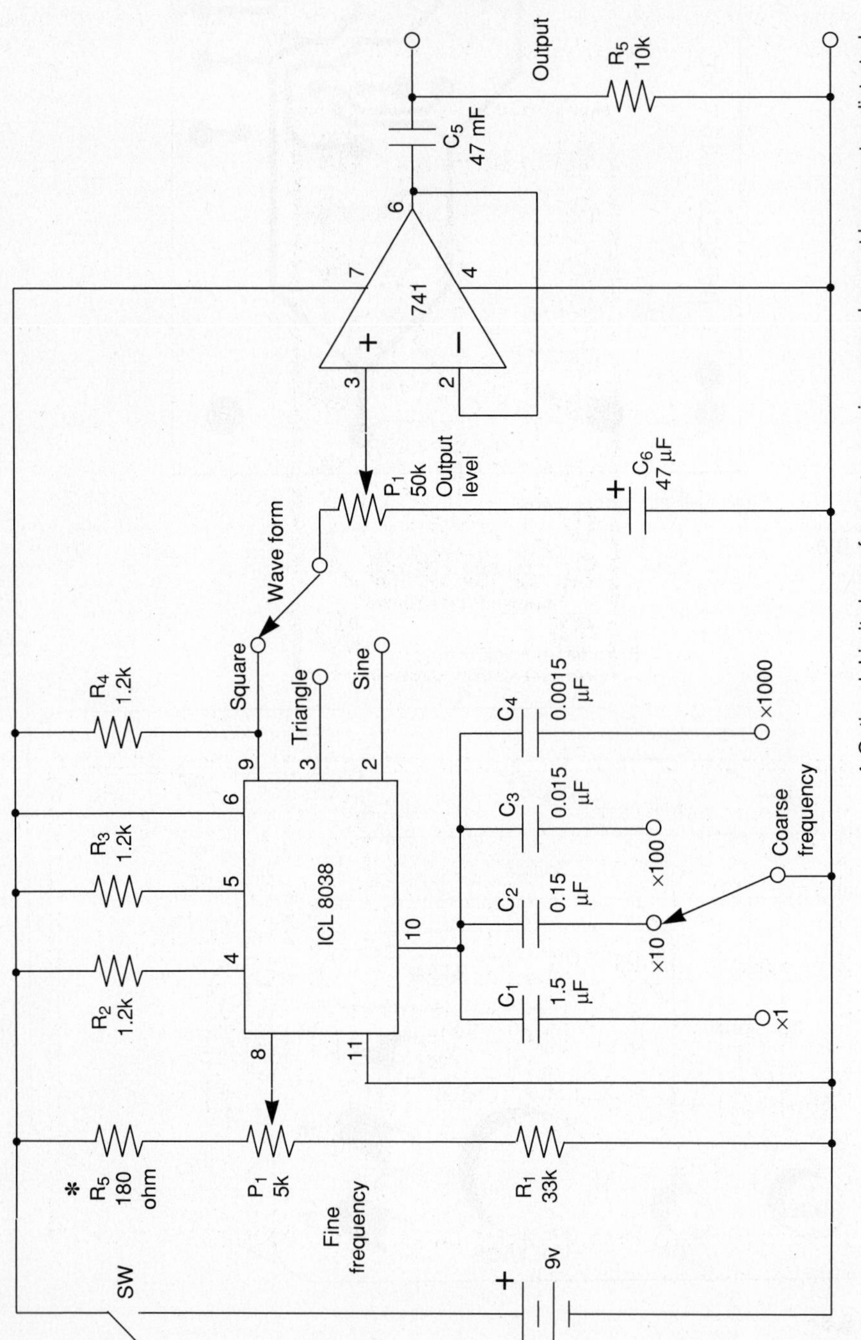

* Optional; Limits lower frequency so sine wave does not become too distorted.

Figure B.9

buffer amplifier (741 operational amplifier) can be added. All resistors are ¼ watt. Controls are fine and course frequency, output wave shape, output amplitude, on/off switch and output terminals/connector. Power is from a 9V radio battery.

Figure B.9 shows the circuit diagram, Figure B.10 the board layout while Figure B.11 the placement of components on the board. Any suitable case can be used, the main criterion being sufficient front panel area to take all the controls.

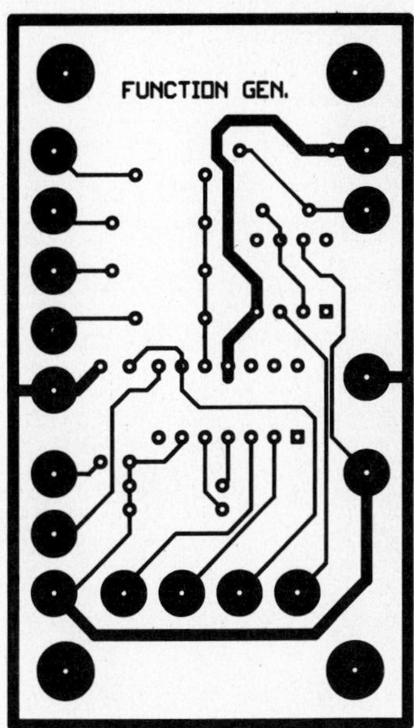

Figure B.10

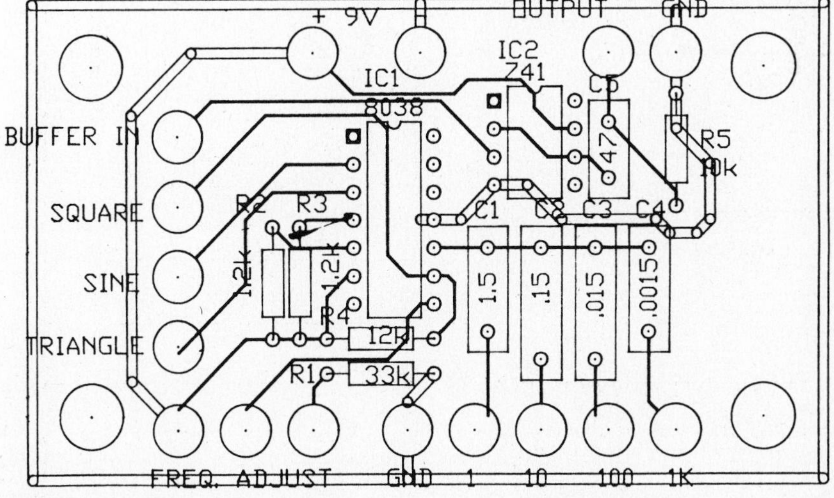

MRHOSC.PCB Check Plot

Figure B.11

Index